Praise for

COLD, HUNGRY AND IN THE DARK

Bill Powers cuts through the deceptions of the natural gas industry's PR machine with a little bit of history—we've been here before!—and a meticulously researched survey of the current disturbing trends in North American natural gas production. Powers' step-by-step dismantling of the abundance myth ought to alarm policymakers, corporate managers, investors, business owners and concerned citizens alike. That alarm needs to translate into action along lines he suggests in order to meet the very real and daunting energy challenges we face.

> —Kurt Cobb, author, *Prelude (a peak oil novel)*, and
> frequent contributor to *The Christian Science Monitor*

Everyone should be aware of Bill's work because it is going to change nearly all current expectations about energy in the US if he is correct.

> —Jim Rogers, author, *Street Smarts: Adventures on the Road and in the Markets;*
> *Investment Biker; A Bull in China; Hot Commodities;* and *Adventure Capitalist*

As our master resources of oil and gas dwindle and get much less cheap, the master wish in a wishful society is the wish to keep driving to WalMart forever. Bill Powers disperses the smoke and shatters the mirrors of a nation wishing itself to death. He lays out in lucid detail why shale gas is not the "game-changer" touted in Wall Street's propaganda mills and he is brave enough to draw conclusions that Americans desperately need to hear, if we're to have any chance of passing successfully through the great bottleneck of history that looms just ahead.

> —James Howard Kunstler, author, *The Long Emergency* and *Too Much Magic,*
> and the *World Made By Hand* novels

Bill Powers has an outstanding ability to see the patterns within a large diverse set of data. He has made it clear why natural gas in North America will experience severe supply issues, resulting in much higher prices in the near future. Highly recommended for investors in the natural gas and energy sector.

> —Harley Kempthorne, P. Eng.

Cold, Hungry and in the Dark is a deep dive into the data on shale gas, and a must-read for anyone involved in energy policy, journalism, or investing. Energy data nerds will love this book. Powers' retelling of the history of natural gas production and policy provides a very useful context for understanding where we are in the progression of gas development, and his careful debunking of the oversold expectations for gas is timely and important. There is no doubt that horizontal drilling and hydrofracking of gas-bearing shales is a triumph of technology that brought us a new abundance of gas in a remarkably short time, but we should heed his warning that it could have a much shorter life than is generally believed. The future of shale gas is uncertain, and this study offers some immensely useful signposts that have been lacking until now.

—Chris Nelder, author, *Profit from the Peak* and *Investing in Renewable Energy*
and editor, www.getreallist.com energy blog

The rhetoric around "100 years of gas" and growing production from shale with low prices for decades is pervasive, and is setting the US economy up for an energy shock that is completely off the political radar. Bill Powers brings a crystal clear focus to the realities of future gas production and prices, and dissects the pervasive hype. It is a must read for understanding the risks of accepting conventional wisdom of eternal cheap energy on long term US energy security. Gas will be a very important input to US energy for the foreseeable future, but assuming it will be cheap and abundant ad infinitum is unrealistic and dangerous in terms of energy policy development – this book provides the facts on future gas supply and prices that are critical in designing a more sustainable energy future.

—Dave Hughes, Fellow, Post Carbon Institute,
and Geologist Emeritus of the Geological Survey of Canada

EXPLODING THE
NATURAL GAS SUPPLY MYTH

COLD, HUNGRY AND IN THE DARK

BILL POWERS

FOREWORD BY ART BERMAN

new society
PUBLISHERS

Inquiries regarding requests to reprint all or part of *Cold, Hungry and in the Dark* should be addressed to New Society Publishers at the address below.

To order directly from the publishers, please call toll-free (North America) 1-800-567-6772, or order online at www.newsociety.com

Any other inquiries can be directed by mail to:

New Society Publishers
P.O. Box 189, Gabriola Island, BC V0R 1X0, Canada
(250) 247-9737

LIBRARY AND ARCHIVES CANADA CATALOGUING IN PUBLICATION

Powers, Bill, 1970–
Cold, hungry and in the dark : exploding the natural
gas supply myth / Bill Powers ; foreword by Art Berman.

Includes bibliographical references and index.
ISBN 978-0-86571-743-5

1. Natural gas reserves — Economic aspects — United
States.   2. Natural gas — Prices — United States.   I. Title.

HD9581.U52P69 2013   333.8'2330973   C2013-901315-6

New Society Publishers' mission is to publish books that contribute in fundamental ways to building an ecologically sustainable and just society, and to do so with the least possible impact on the environment, in a manner that models this vision. We are committed to doing this not just through education, but through action. The interior pages of our bound books are printed on Forest Stewardship Council®-registered acid-free paper that is **100% post-consumer recycled** (100% old growth forest-free), processed chlorine free, and printed with vegetable-based, low-VOC inks, with covers produced using FSC®-registered stock. New Society also works to reduce its carbon footprint, and purchases carbon offsets based on an annual audit to ensure a carbon neutral footprint. For further information, or to browse our full list of books and purchase securely, visit our website at: www.newsociety.com

To Traci and Grace.

Thank you for all of your love and support.

You have made me the luckiest man alive.

Contents

Acknowledgments

Writing a book is a very collaborative effort and I have received wonderful assistance and advice from dozens of great people. For this I am truly blessed. My interactions with everyone involved over the three-year life of this project have both enlightened and challenged me in so many different ways. I hope those who contributed to this project understand how truly grateful I am for their assistance. Thank You.

Below are just a handful of the people who helped get this book published. There has never been a more patient, professional and creative editor/writing coach than Kate Ancell. Without your prodding to punch things up, this book would never have seen the light of day. Additionally, a special thank you to all those who read part of the manuscript and provided feedback. These truly great professionals include Ron Forth (who vastly improved my knowledge of decline curve analysis), Tony van Winkoop, Harley Kempthorne, Robert Carington, Bill Powers of San Diego, CA (who is the "real" Bill Powers and was very helpful in getting me up to speed on alternative and distributed power) and Kevin Haggard. I cannot say "thank you" enough to Art Berman, who graciously let me include much of his groundbreaking work on shale gas and decline curves. Also, I owe a huge debt of gratitude to fellow contrarian and author James Howard Kunstler, who went out of his way to help me get published and put me in touch with the great people at New Society Publishers. It has been a joy to work with Ingrid Witvoet, EJ and Sara at New Society to get this book out the door. Thank you for taking a chance on me. Scott Steedman made much magic happen in taking my raw manuscript and turning it into a real book. Heidi Jo Brady did a wonderful job in taking the photo for the jacket and website.

Lastly, thank you to my family. Thanks to Mom and Dad for all the love and support and the great education you provided me with. Thanks to Jay, Jimmy and Helen for being the best siblings in the world. Finally, I owe a very special thank you to my wife Traci and my daughter Grace.

Traci spent countless hours finding and documenting sources and provided an unending stream of encouragement when the light at the end of the tunnel was barely flickering. Thanks for making this book happen and everything else you do for me. Grace, thank you for your understanding and patience during the many hours I spent writing and researching this book. You will always be my little girl.

Commonly Used Abbreviations

bbl	barrel
bbl/d	barrels per day
boe	barrels of oil equivalent
boe/d	barrels of oil equivalent per day
bop/d	barrels of oil per day
mboe	thousand barrels of oil equivalent
mcf	thousand cubic feet
mcf/d	thousand cubic feet per day
mmcf	million cubic feet
mmcf/d	million cubic feet per day
bcf	billion cubic feet
bcf/d	billion cubic feet per day
tcf	trillion cubic feet
6:1	boe conversion ratio of six mcf to one bbl

Check Out Receipt

Port Orchard
360-876-2224
http://www.krl.org

Friday, May 1, 2015 1:23:5 PM

Item: 39068026516080
Title: Cold, hungry and in the dark : expl
oding the natural gas supply myth
Call no.: 333.8233 POWERS
Material: Book
Due: 05/22/2015

Total items: 1

Find inspiration at your library.
Dream more. Learn more. Do more. Be more.
Kitsap Regional Library

Foreword

by ART BERMAN

When something sounds too good to be true, it probably is. This is particularly true about shale gas. Shale gas is a commercial failure. That is not what the exploration and production companies that produce the gas or the mainstream media and sell-side brokerage companies that help promote the plays tell the public.

Over the past 5 years, I have evaluated, published and spoken about shale gas plays. I am a petroleum geologist and I make my living evaluating prospects and plays based on fundamental geology and economics. Shale gas does not pass the test.

I have written about a phenomenon that I call "magical thinking." Magical thinking focuses on gas production volumes but does not consider cost. This is its catechism: because the volume of shale gas production is great, it must, therefore, be a commercial success; companies would not be producing shale gas if there were no profit is the assumption.

We have lived through a similar situation with real estate. Companies like Merrill Lynch, Bear Stearns and Lehman Brothers would not have been involved in real estate investments that were not profitable. Those companies no longer exist. So much for magical thinking.

Bill Powers has written a fact-based book on the natural gas industry. *Cold, Hungry and in the Dark* gives a detailed account of both the truth and fantasy that surround the shale gas industry and provides a realistic estimate of future gas supply. He makes a clear case that shale gas will never live up to expectations. This important book could not have come at a better time.

Domestic natural gas exploration and production companies have a long history of promising much and delivering less. Their approach has been "trust us" but their performance says that we should not. The shale gas revolution is the most recent chapter in an epic of misrepresentation and disappointment.

Improvements in existing technologies such as horizontal drilling and hydraulic fracturing have played an important role in unlocking substantial supplies of shale gas. The shale players and their advocates emphasize the great volumes of natural gas that have been produced but rarely talk about cost or the sustainability of supply.

A casual investigation of balance sheets and public filings reveals that companies that focus on shale gas ventures rarely have any retained earnings or shareholder value. The flight from gas-weighted enterprises to more liquids-rich plays by shale players in recent years proves that shale gas does not make commercial sense.

The emphasis on shale gas and oil plays does not indicate any great new opportunity. Rather, it reflects that more commercially attractive ventures have been exhausted. Shale plays are not a revolution but, rather, a retirement party as the world exhausts its oil and gas resources.

Cold, Hungry and in the Dark explains many of these limits and limitations in straightforward, non-technical language. Despite the increase in shale gas production, offshore and conventional gas still account for most US production, and these are in decline. By reviewing production trends from five of the largest gas producing states as well as the Gulf of Mexico, America's gas production capacity becomes clear. Unlike the technically recoverable shale gas resource estimates so often mentioned in the mainstream press, Bill's book examines each of the major shale plays and uses production history and well-cited sources to make realistic estimates of future shale gas production. More importantly, readers are taken beyond the numbers to a fact-based examination of the geological reasons behind this reality.

Government agencies and financial sector analysts have consistently gotten energy wrong, and continue to forecast unrealistic future supply and prices for natural gas. This should not be a surprise since few of these people have ever worked a day in the oil and gas industry.

I am in my 35th year as an industry professional and do not pretend to know the right answer about the future of energy and natural gas. I am certain, however, that the capital costs and the enormous number of wells required to maintain present oil and gas supply cannot be sustained except at much higher prices. The tolerance of the public for the environmental impact of so much drilling presents another major uncertainty.

Cold, Hungry and in the Dark provides a critical evaluation of a future in which America moves to higher gas prices and less supply. It examines and explodes many popular misconceptions about natural gas including why its export in meaningful volumes is unlikely and why its widespread use as a transport fuel is a fantasy that ignores the decades required for such massive equipment change and its hidden infrastructure costs. Bill Powers takes a brave stand in this book to expose the truth about the shale promotors who gamble with America's future to promote their own.

— Arthur E. Berman

Preface

"History does not repeat itself, but it does rhyme."
— Attributed to Mark Twain

How many times in the past fifteen years has conventional wisdom been turned on its head? In the last two years of the 1990s, financial commentators and accepted experts across America touted the virtues of technology shares, only to see the internet/technology market implode. More recently, conventional wisdom held that "housing was the best investment a family could make" and that one "should buy all the house possible," since housing prices had not gone down since the Great Depression. At the height of the housing boom, David Lereah, the Chief Economist for the National Association of Realtors, discussed the virtues of owning real estate in his 2006 book *Why the Real Estate Boom Will Not Bust and How You Can Profit From It.*[1] To be fair, Mr. Lereah had plenty of company in getting the housing debacle terribly wrong. In March 2007, Federal Reserve Chairman Ben Bernanke testified before Congress that, "the impact on the broader economy and financial markets of the problems in the subprime markets seems likely to be contained."[2] For someone with superior access to information about our economy and banking system, it is beyond me how Mr. Bernanke could have found the excess leverage and subprime activity to be "contained," when it nearly crashed the world's financial system later that same year.

But surely the experts will not be wrong about the United States' supplies of natural gas—right? Surely—just as we all "knew" that real estate was the be-all, end-all of investments—we now "know" that America really does have a hundred-year supply of natural gas. Surely the industry professionals and government officials who are promoting accelerated use of natural gas as a bridge fuel are relying upon facts?

In reality, the facts—simple facts—do not support the thesis that the US has anywhere close to a hundred-year supply of natural gas. After reviewing America's sources of natural gas supply for more than three years, it is quite clear to me that various groups have vastly overestimated our country's likely natural gas reserves and potential resources. More importantly, America is nearly certain to experience a severe deliverability crisis between 2013 and 2015, due to the natural gas industry's inability to balance supply with growing demand. And when natural gas runs out, America shuts down. We depend on natural gas to produce fertilizer for the majority of food we eat, to heat 51 percent of our homes and to generate 31 percent of our electricity.[3] In other words, without adequate natural gas supplies, the quality of life we have enjoyed for decades falls off a cliff.

So what can we do? The good news is there is lots we can do to mitigate the impact of the coming decline in US natural gas supplies. But before any solutions are put in place, much of today's conventional wisdom about natural gas, its supply and its deliverability will have to be discredited. In this book I will review the history of the natural gas crisis of the 1970s and why it is relevant to the coming crisis; provide a realistic assessment of America's natural gas supply/demand balance, including a detailed review of all of our country's major sources of supply; and suggest solutions to mitigate the effects of the coming natural gas crisis that quietly lurks over the horizon.

The past five years have provided an excellent example of the boom-bust cycle that has forever plagued the natural gas industry. Take the 2008 surge in drilling, for example, which was an attempt to mitigate rising natural gas prices. As a result of this huge drilling increase, America was left with a glut of natural gas supply. Couple this increased supply with the sudden global economic meltdown and recession, which made buyers scarce (no one was buying anything they didn't have to), and prices for natural gas plummeted as producers were forced to offload gas into a weak market. It has taken four years, and a large reduction in natural gas-directed drilling, to return US natural gas storage to more historical levels. Meanwhile, persistently low natural gas prices, at a time of rebounding oil and other commodity prices, have given rise to the widespread, and totally incorrect, belief that the supply outlook for natural gas has changed in a fundamental way. Much of this misconception can be attributed to the idea that advances in technology—specifically, horizontal drilling and hydraulic fracturing—will unlock decades worth of supplies from previously unavailable shale deposits. And, while it is true that shale gas is and will continue to be an important source of supply, the simple fact is that it will have a far smaller impact on future supplies than many energy pundits would have you believe. In fact, the erroneous conviction that advances in technology will keep natural gas cheap and available for the foreseeable future is certain to make the fallout from the coming natural gas deliverability crisis far more severe than it would be if we were prepared for the inevitable.

Another factor that will exacerbate the coming crisis is the misplaced belief that higher natural gas prices always lead to increased production. While it is correct that, in general, higher prices tend to lead to increased supplies, it is also—always—true that hard geological limits always trump squishy economic rules. For example, in 1970—a time of single-digit oil prices—the Cook Inlet area of southern Alaska produced approximately 230,000 barrels of oil per day (bop/d); now, it produces approximately 11,000 bop/d.[4] Despite all the advances in seismic and drilling technologies and the triple-digit oil prices of recent years, the truth is that southern Alaska has likely reached its geological limit for oil production. Similarly, there is strong evidence, which I will provide in Part Three, that the next surge in natural gas prices will not increase supplies, as the entire US has likely reached its geological limit for gas production.

Over the past fifteen years, industry leaders and policy makers have repeatedly identified natural gas as a clean-burning alternative to coal and oil, touted it as a significant source of American jobs and suggested that we find new uses for this supposedly abundant resource. I find several flaws with this line of thinking. First, natural gas is a finite and depleting commodity. It cannot be planted in the spring and harvested in the fall. Second, the US is an increasingly mature hydrocarbon-producing region and, to the surprise of many, continues to rely on imports for a portion of its annual natural gas supply.[5] I find it exceptionally poor economic reasoning to encourage the increased usage of a commodity when current demand cannot be met by domestic supplies.

Unlike believers in the myth that America has a hundred-year supply of natural gas, I am far less sanguine on the outlook for natural gas production. Domestic oil production peaked in 1970–71 and marketed US natural gas production reached a pre-2011 peak in 1973, at 22.65 trillion cubic feet (tcf).[6] While the date of US peak oil production is not in dispute, many industry executives and policymakers believe that the upturn in gas production that began in 2008 is only a harbinger of the bright future of US domestic natural gas deliverability. I disagree. I do not believe the all-time 2011 peak for marketed gas production will ever be materially exceeded, due to the advanced maturity of the majority of America's conventional natural gas fields and limited future shale gas production growth.

How this book is organized

I have divided this book into four parts. In Part One, I will provide a brief history of the US natural gas market and its regulation. I will pay special attention to the gas crisis of the 1970s, the nation's first. Prior to the deregulation of the natural gas market that began in 1978, Washington D.C. bureaucrats set natural gas prices using a byzantine system. I will discuss how these price controls led to severe and long-lasting market imbalances. I will also detail the economic and personal fallout from curtailments of natural gas that occurred in virtually every region of the country. As a result of gas supplies that could not meet demand in the 1970s, businesses lost profits and closed facilities, municipalities closed schools and homeowners watched helplessly as their heating bills skyrocketed. Finally in Part One, I will examine the hundred-year supply myth and how a handful of shale gas promoters have grossly distorted the potential impact of shale gas on America's future supply.

In Part Two, I will review the major sources of domestic demand for natural gas. The major driver of demand since the mid-1990s—similar to the period leading up to the natural gas crisis of the 1970s—has been the building boom of natural gas-fired power plants. Increased consumption of natural gas by the electricity generation sector has made US demand more inelastic than at any point in history. Some of this increased demand has been offset by declining consumption from manufacturers—such as the chemical and fertilizer industries—that have relocated to foreign jurisdictions with lower natural gas prices and labor costs. Apart from the workers in these industries, who have had to find other means of employment, US consumers had the best of both worlds: no local pollution and cheap prices for the products that they consumed during America's recently ended twenty-year shopping spree. However, I see the multi-decade trend of off-shoring industrial production reversing, since triple-digit oil prices and overseas wage inflation are translating into rapidly rising import prices. In other words, the world is about to get a lot less flat in the very near future. Lastly in Part Two, I will briefly discuss the state of the natural gas vehicle (NGV) market in the US and the likelihood of increased demand for natural gas from the transportation sector.

Part Three will review the sources of US natural gas supply. I will focus on the major sources of US production, both conventional and unconventional, and their deliverability potential. To provide as granular a view as possible of US production, I will cover five of the major gas-producing states and the Gulf of Mexico in substantial detail using historical production data. The environmental concerns surrounding hydraulic fracturing will be reviewed in Part Three.

The importance of liquefied natural gas (LNG) to America's overall supply will be examined in this section. Despite the building of numerous re-gasification facilities and the expansion of America's four legacy terminals, there is mounting evidence that growth in LNG imports will be extremely difficult, maybe even impossible. In Part Three, I will also discuss the importance of Canadian natural gas exports to the US market. A very mature Western Canadian Sedimentary Basin (WCSB) and the slow development of new discoveries, such as the Horn River Basin, will end the era of cheap Canadian supply.

In the final part of the book, I will provide suggestions to help reverse the coming crisis and increase transparency in the natural gas market. I will provide numerous solutions that will reduce our dependence on natural gas. Finally in Part Four, I will examine the faulty data collection methods of the US Department of Energy's Energy Information Agency (EIA) and how the agency has consistently overstated the importance of shale gas.

The US must embrace increased efficiency, conservation and renewables immediately or untold and irreversible damage will be done to our economy. We simply have no other choice. America's natural gas resources are far less than many industry leaders and policymakers would like you to believe and natural gas prices are headed far higher than previously thought possible. If America does not reduce its consumption of natural gas it will shortly find itself cold, hungry, unemployed and in the dark. Not a great way to go through life.

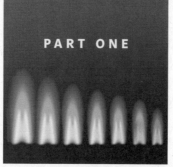

PART ONE

1970S GAS CRISIS SET FOR REPLAY

CHAPTER 1

Early Regulation, Price Fixing and the 1970s Gas Crisis

THE SERIES OF natural gas price spikes and crises in the United States over the past forty years have been both costly and largely avoidable. To get a better understanding of the natural gas crisis that is likely to unfold between 2013–15, let's examine the history of natural gas policy and regulation.

America's first natural gas pipeline systems, built during the latter part of the nineteenth century, carried gas from the early fields in Pennsylvania and West Virginia to major cities on the East Coast. During the 1920s and 1930s, the discovery of enormous deposits in the Texas Panhandle and the Mid-Continent area of Kansas and Oklahoma encouraged pipeline companies to connect supply with growing demand in Midwestern industrial centers such as Chicago. Major advances in technology—such as the oxyacetylene torch and electric arc welding— allowed for the assemblage of the large-diameter pipelines necessary for long-distance gas transmission.[1]

In the early years, it was not economically feasible to have more than a few pipelines in operation, and the barriers to entry were high. As a result, there were few ways for gas to get from the production point to the end consumer—and over time, complaints mounted. The Federal Trade Commission (FTC), a federal agency formed in 1914 whose mission

[handwritten margin note: late 1800s]

3

includes the promotion of consumer protection and the elimination or prevention of anti-competitive business practices, began investigating.[2] The FTC concluded that four of the major pipeline companies (which controlled over 55 percent of all gas lines operated in the US) represented a natural monopoly in their capacity to sell and transport gas.[3] Those same firms were also the sole buyers of gas at the wellhead (natural monopsonies).[4] The FTC found the typical conduct that often arises when monopolies/monopsonies exist—price gouging, significant profit making by the pipeline companies at the expense of customers and poor service in transporting gas.[5] There was also concern that the more competitive area of the gas industry, exploration and production, would also be affected by the potentially unfair business practices of the pipeline companies.[6] In 1935, the FTC issued a report that recommended federal regulation of the pipelines.[7]

In 1938, the Natural Gas Act (NGA) was born. The act gave the Federal Power Commission (FPC), created in 1920 as part of the Federal Water Power Act, jurisdiction over regulation of interstate natural gas sales.[8] Prior to that time, state regulators were charged with oversight of the natural gas industry within their respective states. The NGA required that pipeline companies apply only "just and reasonable" fees to customers. In addition to pipeline fees, the NGA also granted the FPC control of the construction of interstate natural gas pipelines but not authority to regulate natural gas prices at the wellhead.

Until the mid-1950s, wellhead natural gas prices remained in check, as producers had little leverage when negotiating contracts with pipeline companies. US wellhead natural gas prices averaged only $.055 per mcf during the 1940s, and, in the first five years of the 1950s, only $.08 per mcf.[9] But the post-WWII economic boom—and surging demand for natural gas—finally gave producers the bargaining power to demand price increases from the pipeline companies. It was at this time that several US Supreme Court cases determined that wellhead prices were also subject to federal oversight, if, and only if, the selling producer and the purchasing pipeline were affiliated companies. The FPC contended that, if a producer sold gas into an affiliated pipeline, they would have too much control over prices in the area the pipeline served. Interestingly, the FPC put no restrictions on producers who sold their gas into unaffiliated pipelines. In an effort to sell gas at higher prices, many utility companies simply spun off their natural gas-producing divisions into stand-alone companies.[10]

Then in 1954, the Supreme Court case of Phillips Petroleum vs. State of Wisconsin dramatically altered the course of the natural gas industry. The court ruled that natural gas-producing companies that sold into interstate pipelines were considered "natural gas companies" under the Natural Gas Act, whether or not they sold their gas into an unaffiliated pipeline. Consequently, they were now subject to oversight by the FPC. The Supreme Court's decision meant that the FPC would now set wellhead gas prices that allowed the companies to cover the cost of producing gas, plus a "fair" profit.[11] In order to set a fair price for natural gas at the wellhead, the FPC initially undertook the unwieldy approach of treating every producing company as its own separate public utility. With this decision, each producing company received a price for its gas based on the cost of its production. Needless to say, this created an administrative nightmare for the FPC, because the thousands of producing companies all had their own, varied cost structures.

By 1960, after finally realizing that setting different rates for each producing company was impractical, the FPC decided to set wellhead prices on a geographic basis: it divided the country into five areas and set an interim ceiling for natural gas prices for each area, based on its average sales contract for the period 1959–60.[12] The FPC intended these price ceilings to be temporary, until it could determine a "just and reasonable" formula for each area based on producer costs. However, this new strategy for setting prices also proved very difficult to implement, because the various producers within each geographic region had different cost structures. For example, companies that operated smaller fields—or were producing from deeper or more technically challenging horizons where drilling and/or operating costs were higher—were at a distinct disadvantage to producers operating large, low-cost fields. By 1970, new rates for only two of the five producing regions had been established. More importantly, rates were frozen at 1959 levels for three of the five producing regions, making natural gas production only marginally profitable for many producers. But this low pricing and seemingly limitless supply was a boon for American consumers: between 1949 and 1970, due both to the booming postwar economy and prices that were largely locked at 1959 levels, America's appetite for natural gas exploded.[13] Annual marketed natural gas production during this halcyon period of American manufacturing jumped an enormous 304 percent, from 5.42 trillion cubic feet (tcf) per annum in 1949 to 21.92 in 1970.[14]

The first signs that demand for natural gas was outstripping supply appeared in 1970, when industrial users found their supplies curtailed to avoid cutoffs to residential consumers.[15] A study by the US government's Office of Technology Assessment (OTA), published in 1975, detailed the scope of the curtailments in the early 1970s and projected a further deficiency during the winter of 1975–76.[16] Table 1.01 below shows the cutoffs to industrial users in the early 1970s.[17]

While the OTA's study is helpful in understanding the amount of gas that was unavailable to customers in the early 1970s, it doesn't even scratch the surface of the damage that resulted from natural gas curtailments, both to the American economy and to the American psyche.

By early 1970, large industrial concerns such as US Steel were experiencing natural gas curtailments. As part of the company's earnings release for the first quarter of that year, Chairman Ed Gott attributed US Steel's drop in profits to "difficult and costly operating conditions and the curtailment of operations in many plants due to natural gas shortages."[18] In 1973, the University of Texas was forced to cut off heat to student dormitories in Austin and delay the start of its second semester by a week.[19] In November 1975, textile manufacturer Dan River Inc., based

TABLE 1.01. Total curtailment volumes of interstate pipelines.

Year (April 1 to March 31)	Curtailment (tcf)	Approximate percentage of total consumption
1971–72	.48	2.18%
1972–73	.82	3.17%
1973–74	1.19	5.50%
1974–75	2.01	9.86%
1975–76	2.92*	14.79%

Source: Congressional Board, Office of Technology Assessment document, EIA.
* Projected.

in Danville, Virginia, lost half of its gas supply after pipeline company TransCo reduced natural gas deliveries to the city by 67 percent, temporarily cutting off gas supplies to many industrial users. Dan River was able to operate at half its normal capacity due to an emergency program that allowed them to purchase gas directly from producers at prices that were double the $.90 per mcf it was paying to TransCo.[20] And in February 1977, approximately two million students—and thousands of teachers—in Ohio were left idle for several weeks when their schools were closed due to lack of natural gas supplies.[21]

One of the long-forgotten aspects of the 1970s gas crisis was the complete lack of transparency in reserve estimation. It was not until 1978 that the Securities and Exchange Commission required oil and gas companies to report their reserves in their annual financial filings. With America's publicly traded producers, which also happened to be the largest producers, not providing reserve reports to their shareholders or to the government on a regular basis, the world was largely in the dark about the true size of America's natural gas reserves. Industry's message to the government and consumers was "trust us."

In the days before the creation of the Department of Energy, the FPC relied on the industry-sponsored American Gas Association (AGA) to provide reserve data to assist it in setting natural gas prices. Needless to say, this reliance on industry-provided data to determine its own profitability was fraught with conflicts of interest. By 1975, the Federal Trade Commission's (FTC) Bureau of Competition found that the AGA had lowballed its yearly reserve estimates by 24 percent in an effort to win rate increases for its members.[22]

Another very important, though often forgotten, takeaway from the 1970s natural gas crisis is that higher prices do not always result in more production—though this was certainly not the message of industry cheerleader AGA. Indeed, there was no organization more vocal than the AGA in advocating that America's ample reserves of natural gas—and higher natural gas prices—would bring about more production. Figure 1.01 is an example, from *Life* magazine in 1971, of the AGA's campaign to spread the (incorrect) idea that raising prices would solve the natural gas crisis.[23]

Our country's gas supplies:

What the gas industry is doing to be sure your home has enough gas.

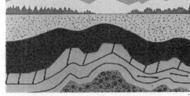

The gas supplies are there. The problem is getting at them. Today our country faces a growing need for all types of energy, including gas. Our continent has gas. Huge reserves of it. But much of it is deep down. Hard to get at. Some of it is under water. And there have been important new discoveries in the far-off Arctic. It's going to take time and money to make additional gas available.

There's no worry your home will run out of gas. In some areas, the amount of additional natural gas to large-scale industrial users is limited. But wherever there is any problem, gas companies are giving top priority to their residential customers. We've been serving you for a hundred years – and we don't intend to stop now.

The gas industry is drilling, piping, importing, researching to increase the gas supply. The gas industry and government are working together on an accelerated research program to convert coal into clean-burning gas. The gas industry is piping in gas from Canada and importing liquefied natural gas from overseas.

It will take higher prices to keep gas coming. For years gas prices at the wellhead have been kept artificially low – while drilling and other costs have skyrocketed. Recently more realistic price levels have been approved to get the huge job of exploration and drilling done. The higher costs incurred will mean somewhat higher prices to you, but gas will still remain more economical comparatively than other forms of energy. And it's worth more to keep this essential energy coming.

It make sense to save clean gas energy. Nowadays we all know we should use our natural resources wisely. There are many things we can do to save natural gas – like weatherproofing our homes and not wasting gas when we cook. Saving gas makes sense even after new supplies become available. Gas is clean energy – a pure, natural energy that doesn't foul the air we breathe. It's going to be important in giving us a cleaner world to live in. Natural gas is valuable. Use it wisely.

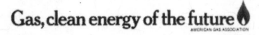

Gas, clean energy of the future 🔥

AMERICAN GAS ASSOCIATION

FIGURE 1.01. American Gas Association advertisement, *Life* magazine, October 22, 1971.

Ironically, despite this protestation from the AGA—that "it will take higher prices to keep gas coming"—higher natural gas prices did not, in fact, keep the gas coming—at all. Less than two years after that ad appeared, US-marketed natural gas production peaked at 22.64 tcf per annum. More importantly, *despite a more-than-400-percent increase in wellhead prices between 1971 and 1978, marketed US natural gas production declined 11.2 percent during this period.* Table 1.02 clearly shows how increased wellhead prices for natural gas did not result in increased production.

Apparently, the AGA's promise that gas would be available at the right price was absolute balderdash. As I will discuss in Chapter 4, I find many similarities between the AGA's campaign to create a false sense of security about America's natural gas deliverability in the 1970s and more recent campaigns to generate complacency with regards to America's future natural gas deliverability.

The gas crisis of the early 1970s had a significant impact on the electric utility industry. Since America was no longer able to increase its supplies of natural gas thereafter, the only way to rebalance the nation's electricity market was to destroy demand *for* natural gas. It is difficult to overstate the importance the growth in nuclear and coal electricity-generating capacity had in mitigating the crushing tightness in the

TABLE 1.02. Marketed US natural gas production and average wellhead price.[24]

Year	Marketed US natural gas production in tcf	Average wellhead price per mcf
1968	19.32	$.16
1969	20.69	$.17
1970	21.92	$.17
1971	22.49	$.18
1972	22.53	$.19
1973	22.64	$.22
1974	21.60	$.30
1975	20.10	$.44
1976	19.95	$.58
1977	20.02	$.79
1978	19.97	$.91

Source: eia.doe.gov/dnav/ng/hist/n9050us2a.htm, eia.gov/dnav/ng/hist/n9190us3A.htm.

natural gas market during the 1970s. Coal and nuclear electricity generation grew 65 percent and 1,051 percent respectively from 1970 through 1980 and allowed for the percentage of electricity generated from natural gas to drop from 24.3 percent in 1970 to only 15 percent in 1980. To meet increased demand for electricity now that natural gas was becoming increasingly expensive and scarce, America went on the largest coal-fired power plant building binge in its history. Figure 1.02 shows the enormous coal generation capacity brought online during this period.[25]

As can be seen from Table 1.03, natural gas declined as a feedstock for electricity generation while the use of alternative fuels exploded during the 1970s.[26]

The bad news for us, now? Such a reduction in our reliance on natural gas for electricity generation is highly unlikely during the upcoming natural gas deliverability crisis. Why? Simply put, we're not putting our policies where our problems are: instead of reducing our reliance on natural gas as a feedstock ahead of the deliverability crisis, the US continues to do the exact opposite. The EPA's MATS (Mercury and Toxic Air Standard) rules—which are set to take effect in 2015 and will likely accelerate the shutdown of 15 percent of America's coal generation fleet—guarantee that we will rely more on natural gas for power generation over the next five years. The addition of significant new

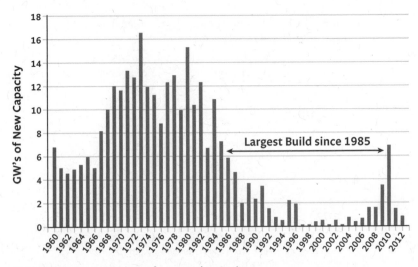

FIGURE 1.02. Last 50 years of new coal capacity

TABLE 1.03. 1970s Electricity Generation by Fuel Type in Billion Kilowatt Hours.

Year	Natural gas	Coal	Petroleum	Nuclear	All sources
1970	372.9	704.4	184.2	21.8	1,535.1
1971	374.0	713.1	220.2	38.1	1,615.9
1972	375.7	771.1	274.3	54.1	1,753.0
1973	340.9	847.7	314.3	83.5	1,864.1
1974	320.1	828.4	300.9	114.0	1,870.3
1975	299.8	852.8	289.1	172.5	1,920.8
1976	294.6	944.4	320.0	191.1	2,040.9
1977	305.5	985.2	358.2	250.9	2,127.4
1978	305.4	975.7	365.1	276.4	2,209.4
1979	329.5	1,075.0	303.5	255.2	2,250.7
1980	346.2	1,161.6	246.0	251.1	2,289.6

Source: EIA eia.doe.gov/emeu/aer/pdf/pages/sec8_8.pdf.

nuclear generation capacity is also unlikely for the foreseeable future, especially since Japan's nuclear accident of March 2011 heightened fears over nuclear safety. Further, while advances in technology have made renewable power sources such as wind and solar much more competitive within the mainstream market, renewables still have a long way to go before they will make a material contribution in meeting America's electricity demands. In 2010, all forms of renewable generation, outside of hydroelectric sources, accounted for *only four percent* of all electricity generated in the US.[27]

But let's get back to the 1970s. This was a time when elected officials wanted to be seen to be "doing something" about the natural gas shortage, though few were willing to tackle reform in any meaningful way. With his presidency crumbling at his feet, Richard Nixon delivered a special message to Congress on energy policy on April 18, 1973, that was steeped in blather and political double-talk about the many dilemmas facing the natural gas market. President Nixon clearly realized that the existing pricing structure of the natural gas industry was no longer working, but was not yet ready to let market forces determine the price of natural gas:

It is clear that the price paid to producers for natural gas in interstate trade must increase if there is to be the needed incentive for increasing supply and reducing inefficient usage. Some have suggested additional regulation to provide new incentives, but we have already seen the pitfalls in this approach. We must regulate less, not more. At the same time, we cannot remove all natural regulations without greatly inflating the price of gas currently in production and generating windfall profits.

— President Richard Nixon, 1973[28]

In June 1974, the FPC finally—belatedly—took the half-measure of abandoning the regional pricing structure for natural gas and set the first national rate for new natural gas contracts, at $.42 per mcf. It should be noted that not all natural gas was priced at this rate; only wells that began selling gas *after* January 1, 1973, were entitled to receive this new price.[29] However, even this action did little to slow down accelerating shortages of natural gas, and production continued to drop. Despite an average national price for *new* natural gas in 1973 that was 91 percent higher than the average wellhead price (see Table 1.02), shortages persisted and industrial users continued to be curtailed. As time went on, natural gas shortages and rising prices continued to sap consumer confidence, increase unemployment and add to America's newfound energy insecurity. *Still*, policy makers lacked the political will to do anything either new or of substance to fix the problem—despite the fact that continued increases in the price for "new gas" were having little to no impact on the crisis. In great frustration, FPC Commissioner Rush Moody resigned in March 1975, angry over Congress' refusal to push through legislation required to alleviate the crisis. In his resignation letter to President Ford, Mr. Moody said:

> I can no longer accept the lack of leadership of the so-called "energy leaders" in Congress who, by their refusal to recognize the failure of natural gas regulation, continue to deceive the American people into the belief that wellhead price regulation serves the public interest. Nor can I accept the view that the Commission must, in the absence of legislative change, do no more than preside over the demise of the interstate natural gas market.
>
> As I am sure you perceive, the gas consumers of this country have been betrayed by the false premise that natural gas can be

supplied indefinitely at rates that will not permit replenishment of the sources of supply. The disruptive effects of the twenty-year Federal effort to make this false premise effective are now being felt in increasing dependence on imported oil, and massive curtailments of natural gas service; tragically, the worst effects of the regulation-induced natural gas shortage are yet to come.[30]

In 1976, the FPC took a bold step and raised the national price ceiling for newly discovered gas to a previously unthinkable $1.42 per mcf (up from only $.42 per mcf in 1974) in an ineffective attempt to put an end to continued supply disruptions.[31] In an effort to be seen as taking a proactive approach to the energy crisis, newly elected President Carter signed into law the Emergency Natural Gas Act of 1977 less than two weeks after taking office. This act gave the chairman of the Federal Power Commission the power to transfer gas from areas of surplus to areas in short supply until April 30, 1977. It also allowed pipeline companies to purchase intrastate gas from producers at "fair and equitable" prices for delivery through the interstate pipeline system until July 31, 1977.[32] When he signed the act, President Carter had the following to say:

I asked the Congress just a few days ago to give emergency powers to me and to the Federal Power Commission and others to provide some reasonable assurance to the American people that natural gas could be placed in our country where it's needed most.

We now have literally thousands of factories closed down, about 500,000 workers out of jobs, because of the natural gas shortage. [Emphasis added][33]

After six decades of ineffectiveness, the Federal Power Commission was finally put out of its misery with the Department of Energy Organization Act of 1977, which created the Department of Energy (DOE) and the natural gas industry's new regulator, the Federal Energy Regulatory Commission (FERC). While some in Washington hoped that the DOE and FERC would bring about more effective regulation of the natural gas industry, it turned out to be a case of "second verse, same as the first."

Despite the efforts of President Carter and the FPC, the natural gas shortage became so acute in January 1977 that the US Mint announced that two facilities, one in New York and one in Philadelphia, would reduce their consumption of natural gas. You know things are bad when the US Mint gets its supply of natural gas curtailed![34]

While the temporary lifting of price limitations on natural gas prevented some of the economic damage the cold winter of 1977 would have otherwise inflicted, additional actions would be needed for the gas market to find a new equilibrium. Despite several years of damaging natural gas supply shortfalls, sweeping legislative reform of the natural gas industry had to wait until 1978, when the National Energy Act (NEA) was passed through Congress and signed into law by President Carter,

FOR IMMEDIATE RELEASE   JANUARY 28, 1977

MINT FACILITIES TO CURTAIL USE OF NATURAL GAS

    Deputy Director of the Mint Frank H. MacDonald announced today that in accordance with requests from the States of Pennsylvania and New York, the U. S. Mint at Philadelphia and the New York Assay Office will stop using natural gas in their production operations effective immediately.

    The Mint facilities, like other industrial operations, have been asked to reduce consumption of natural gas to the lowest possible levels without causing irreparable damage to equipment. The curtailed usage will last for the duration of the natural gas crisis.

    Affected operations will include die manufacturing and coin production in Philadelphia and all melting operations involving natural gas at the New York Assay Office. However, the two facilities will continue to work from inventories during the crisis period.

FIGURE 1.03: Letter from the US Mint.

on November 8, 1978. Within it were five separate acts, each designed to help the country muddle through the worst energy crisis in its history:[35]
- Public Utility Regulatory Policies Act (PURPA) (Pub.L. 95-617)
- Energy Tax Act (Pub.L. 95-618)
- National Energy Conservation Policy Act (NECPA) (Pub.L. 95-619)
- Power Plant and Industrial Fuel Use Act (Pub.L. 95-620)
- Natural Gas Policy Act (Pub.L. 95-621)

The two acts aimed directly at alleviating the worsening natural gas crisis were the Power Plant and Industrial Fuel Use Act (FUA) and the Natural Gas Policy Act (NGPA). The FUA required power plants to convert to coal where feasible, and encouraged reductions in the use of natural gas in utility and industrial boilers.[36] The NGPA increased wellhead price ceilings to provide additional economic incentives for producers. These new price ceilings and the process for increasing rates were set out in statute, rather than relying on an independent body to determine these rates.[37] Additionally, the NGPA allowed for the phasing out of natural gas wellhead price controls for newly drilled wells, with the goal of total deregulation by 1985. However, the act also mandated that gas brought into production before the passage of the NGPA would forever be subject to previous regulations and price limits.[38]

In the next chapter I will explore how the NGPA created a two-tiered pricing structure for natural gas and set the stage for the development of "take or pay" contracts between producers and pipeline companies. I will also examine why natural gas prices continued to rise for six more years after the passage of the NEA.

CHAPTER 2

1978 to 1984: The Failure of Policy Half-Measures

WHETHER THE MARKET is for commodities or widgets, increased prices have historically been able to coax more supply from producers. Unfortunately, reality does not always cooperate with economic theory. Table 2.01 clearly displays how little impact the removal of the price ceiling for new gas had on domestic production. Unbelievably, US marketed production declined 10.58 percent in the six years between 1979 (the first full year after the NGPA was enacted and natural gas prices were partially deregulated) and 1984, despite a 125-percent increase in the average wellhead price.

Part of the reason natural gas production did not rise after the lifting of price controls for new gas was that pipeline companies were caught in a two-tiered pricing structure that eventually threatened their very existence. Specifically, in an effort to secure natural gas supplies in the years immediately following the passage of the NGPA, many pipeline companies entered into "take-or-pay" contracts. Such contracts required them to pay producers for gas whether or not they were able to sell the gas on to consumers. Take-or-pay contract provisions became popular immediately after the NGPA became law in 1978, due to section 107 of the act, which allowed pipeline companies to pay market rates for "high-cost" gas (gas from deep reservoirs). According to the Congressional Budget

TABLE 2.01. Average US Natural Gas Price
and Annual Production 1979–1984.[1]

Year	Wellhead price of NG (mcf)	Marketed NG production (tcf)
1979	$1.18	20,471
1980	$1.59	20,179
1981	$1.98	19,955
1982	$2.46	18,711
1983	$2.59	16,884
1984	$2.66	18,304

Source: Energy Information Agency.

Office (CBO), by June 1982 high-cost gas averaged $7.41 per mcf—*even though the average wellhead price for 1982 was only $2.46; quite a difference!* But pipeline companies paid up in order to secure new supplies, because they believed they would be able to blend high-priced new gas with lower-priced legacy gas and still remain profitable.[2]

According to a 1983 CBO study on the deregulation of the natural gas market, much of the impetus behind the deregulation of prices in 1978 was to allow them to rise enough so that natural gas would achieve parity on an energy equivalent—or British thermal unit (BTU) basis—with oil.[3] (Approximately six thousand cubic feet of natural gas contain the energy equivalent of one barrel of oil.) While the NGPA did succeed in increasing gas production in 1979, the year after its enactment, its two-tiered pricing mechanism caused serious market distortions. The 1981–82 recession, which caused a drop-off in demand and lower prices for "new gas," resulted in substantial economic losses for pipeline companies, who were locked into take-or-pay contracts despite lower-cost new gas being available on the open market. Instead of taking all the losses they would have incurred by selling gas to customers below the price they were paying producers, they simply cut back on purchases of cheaper new gas. According to the CBO, as the price of oil fell and the price of gas continued to rise in the early 1980s, average natural gas prices came close to parity with oil on a BTU basis, and gas even became expensive to burn compared to fuel oil for industrial boilers. Here is how the CBO described the huge distortions caused by take-or-pay contracts in the natural gas market in November 1983:

Gas prices now appear to have risen to levels at which they rival other fuels. This is demonstrated by the fact that the average price found in new contracts for high-cost gas fell by about $1 per thousand cubic feet from June 1981 to June 1982. Moreover, many pipelines are now renegotiating downward the prices they pay to gas producers. These price declines are evidence that demand for relatively higher price gas has fallen and that pipelines can no longer raise the price of gas paid by their customers. As pipelines found themselves unable to sell all their gas, their "take-or-pay" provisions went into effect. Obligated to buy new and more expensive supplies, pipelines were often forced to cut back their purchases of less expensive gas, precisely the opposite of the sequence that would presumably occur in a competitive market.[4]

With the price of de-controlled natural gas above its oil equivalent by the end of 1982, US residential consumers cut back on their consumption of natural gas, as did the manufacturing and electricity-generating sectors. Table 2.02 shows how significantly natural gas usage declined in the decade that followed the NGPA enactment, despite otherwise robust economic growth in the US. Industrial demand for natural gas fell an incredible 15.5 percent between 1978 and 1987 as many industrial concerns switched back to using fuel oil, widely viewed as a more reliable fuel source than natural gas, in their boilers. The electric power industry also reduced its consumption of natural gas, continuing to turn to coal (along with nuclear power) as feedstock to meet growing electricity demand.[5] Despite the lower emissions of natural gas when compared to coal, fears surrounding natural gas supply reliability led to an 11 percent decline in its consumption by the electricity generation sector between 1978 and 1987, while the amount of electricity generated from coal-fired power plants grew 58 percent.[6] During this same period, electricity generated from nuclear power sources in the US rose 10.5 percent, despite the 1979 Three Mile Island nuclear accident and resulting abandonment of many nuclear projects already underway.[7]

The decade between 1978 and 1987 also saw demand from residential and commercial consumers—mostly customers involved in the delivery of non-manufacturing goods and services—drop 12 and 7.6 percent respectively. While some of these decreases can be explained by changes in weather patterns, they are large enough to indicate that many homes and businesses that were once heated with natural gas converted to electric heat.

While wellhead natural gas prices peaked in 1984 at an average price that would not be seen again for another 16 years, the credit for breaking the upward spiral of prices of the late-1970s cannot be assigned to any policy initiatives or increased supply. *Rather, demand destruction eventually re-balanced the US natural gas market.*

One of the most important pieces of legislation spawned by the energy crisis of the 1970s was the Crude Oil Windfall Profit Tax of 1980. Buried deep within a law designed to force the oil industry to pay increased taxes was a clause that gave tax benefits for the development of unconventional oil and gas. These benefits, which came to be known as Section 29 credits due to their location in the IRS code, gave producers of unconventional energy resources such as oil shale, shale gas and coal bed methane (CBM) a credit of $.50 per million BTUs, an amount that eventually reached $1.10 per BTU by 2002 due to inflation adjustments. Section 29 credits were fundamental to the jumpstarting of the US unconventional natural gas industry. But despite their long-term positive effects on the development of America's unconventional resources, they

TABLE 2.02. Natural Gas Consumption in Decade after Passage of the NGPA (in tcf).[8]

Year	Residential	Commercial	Industrial	Elecricity generation	Total	GDP growth
1978	4,903	2,601	8,405	3,188	19,627	5.6%
1979	4,965	2,786	8,398	3,491	20,241	3.1%
1980	4,752	2,611	8,198	3,682	19,877	−0.3%
1981	4,546	2,520	8,055	3,640	19,404	2.5%
1982	4,633	2,606	6,941	3,226	18,001	−1.9%
1983	4,381	2,433	6,621	2,911	16,835	4.5%
1984	4,555	2,524	7,231	3,111	17,951	7.2%
1985	4,433	2,432	6,867	3,044	17,281	4.1%
1986	4,314	2,318	6,502	2,602	16,221	3.5%
1987	4,315	2,430	7,103	2,844	17,211	3.2%

Source: US EIA and US Department of Commerce/BEA.

were not enough to keep average wellhead prices from continuing to rise for four more years after they came into existence.

Economic history has shown that while price controls may "work" (i.e., bring relief to certain groups) for extended periods of time, they eventually distort the forces of supply and demand and almost always result in higher prices and more economic disruption than otherwise would have occurred. As I have detailed in the first two chapters of this book, wellhead price controls distorted natural gas prices for decades before nearly destroying the US economy and adding to the galloping inflation of the late 1970s. While an in-depth discussion of price controls for energy markets or any other market is beyond the scope of this book, the five decades of federal natural gas price controls that began with the Natural Gas Act of 1938 and ended with the Natural Gas Wellhead Decontrol Act of 1989 (which I will discuss in the next chapter) should be seen as *prima facie* evidence that price controls for natural gas simply do not work.

In the next chapter I review the era of natural gas deregulation and the rebound in natural gas demand.

CHAPTER 3

1984 to 2000:
The Era of Deregulation

W HILE THE Natural Gas Policy Act of 1978 was the first step in the deregulation of the market, the Ronald Reagan era marked the beginning of the largely deregulated natural gas market we know today. The first major deregulation push in the 1980s surrounded the separation of the *purchase* of natural gas from the *transportation* of gas through the nation's pipeline system. Under the NGA and the NGPA, pipelines purchased natural gas from producers, transported it to their customers (mostly local distribution companies or LDCs) and sold the bundled product for a regulated price. Instead of being able to purchase the natural gas as one product and the transportation as a separate service, pipeline customers were only offered bundles.[1] In an effort to boost usage of natural gas after many industrial users switched to fuel oil in the late 1970s, several pipeline companies developed special marketing programs (SMPs) for industrial users who could switch between natural gas and alternative fuels. SMP agreements allowed these industrial customers the right to purchase gas directly from producers, and to transport this gas via the interstate pipeline system for a fee.[2] Though SMPs were approved by FERC, they were eventually found to be discriminatory in the courts, as not all consumers were able to take advantage of them.

The issuance of FERC 436 in 1985 established a voluntary framework under which interstate pipelines could act as transporters of natural gas. This order provided all customers the same opportunities and therefore avoided the discrimination problems of the earlier SMPs. FERC allowed pipelines to offer transportation services to customers who requested them on a first-come, first-served basis. There were transportation rate minimums and maximums, but within those boundaries the pipelines were free to offer competitive rates to their customers. Although the framework established by Order 436 was voluntary, all of the major pipeline systems eventually took part.[3] Pipeline customers realized cost savings, in that the spot market price of natural gas was often much lower than the price offered by pipeline companies, who were still bound by many take-or-pay contracts. Since Order 436 allowed customers to purchase natural gas from numerous sources, it became known as FERC "Open Access Order."

Struggling under the enormous losses from their take-or-pay contracts, the pipeline industry received its own version of a bailout in 1987 with FERC Order 500. This encouraged pipeline companies to wind down take-or-pay contracts and allowed a portion of the cost of doing so to be passed along to customers. It also allowed the LDCs, to which a portion of the take-or-pay resolution costs were passed, to further pass these costs along to retail customers, in the form of higher rates.[4]

Another important piece of natural gas-related legislation passed during the 1980s was the 1989 Natural Gas Wellhead Decontrol Act (NGWDA), which finally set the conditions for market forces to establish an equilibrium price for natural gas. Under it, all remaining NGPA price regulations were to be eliminated by January 1, 1993, allowing the market to completely determine the price of natural gas at the wellhead.[5]

FERC Order 636, issued in 1992, required that pipelines separate their transportation and sales services. It made the voluntary open access to transportation that was encouraged in FERC 436 mandatory, so that all pipeline customers would have a choice in selecting their gas sales, transportation and storage services from any provider, in any quantity.[6] The ability of producers and consumers to freely access the interstate pipeline system was fundamental to the development of gas marketing firms such as Enron and the formation of the independent power industry (IPP).

The Gas Bubble Emerges

After natural gas prices finally topped out in 1984, a period of remarkable price stability developed. During this period, which lasted for approximately 15 years (1985 through most of 2000), natural gas prices largely bounced between $1.50 and $2.50 per mcf. When prices fell to unprofitable levels, many producers shut in their uneconomic wells. During periods where prices approached $2.50 per mcf, new and shut-in wells were brought onstream, until a glut of supply developed and prices retreated. While consumers welcomed a decade and a half of stable natural gas prices after the shortages and price volatility of the 1970s, many US natural gas producers considered the era a "gas bubble," since prices were never able to break above materially $2.50 per mcf for any length of time. Additionally, the 282 percent increase in Canadian imports during this 15-year timeframe (see Table 3.01) masked the growing maturity and declining productivity of America's natural gas production base. For ex-

TABLE 3.01. US Supply, Ave. Price, US Rig Counts and Active US Natural Gas Wells (1985–2000).[7]

Year	Marketed US production in Tcf	Canadian imports in Tcf	Average price per mcf in USD	Average US natural gas rig count	Active US natural gas wells
1985	17.27	.926	$2.51	NA	NA
1986	16.86	.749	$1.94	NA	NA
1987	17.43	.993	$1.67	NA	NA
1988	17.92	1.276	$1.69	354	NA
1989	18.10	1.339	$1.69	404	262,483
1990	18.59	1.448	$1.71	463	269,790
1991	18.53	1.710	$1.64	351	276,987
1992	18.71	2.094	$1.74	331	276,014
1993	18.98	2.267	$2.04	364	282,152
1994	19.71	2.566	$1.85	427	291,773
1995	19.51	2.816	$1.55	385	298,541
1996	19.81	2.883	$2.17	465	301,811
1997	19.87	2.899	$2.32	564	310,971
1998	19.96	3.052	$1.96	560	316,929
1999	19.80	3.368	$2.19	496	302,421
2000	20.20	3.544	$3.68	720	341,678

Source: EIA, Baker Hughes Inc.

ample, as shown in Table 3.01, between 1989 and 2000 the active natural gas well count increased 30 percent and the average natural gas directed rig count rose 78 percent, while production increased only 11.6 percent.

No piece of legislation was more important in increasing demand for natural gas in the 1980s than the 1987 Natural Gas Utilization Act, which repealed much of the 1978 Power Plant and Industrial Fuel Use Act. The electricity-generating industry very much welcomed the opportunity to use this low-cost fuel source. This is how the US Energy Information Administration's website describes the big jump in consumption after the passage of the Natural Gas Utilization Act:

> Natural gas consumption for electric generation rose from 2.6 trillion cubic feet (tcf) in 1988 to 5.7 tcf in 2002, an increase of about 119 percent. Natural gas consumption for industrial processing rose from 6.4 tcf in 1988 to 7.6 tcf in 2002, an increase of almost 19 percent. Therefore, total natural gas consumption for electric generation and industrial processing increased approximately 47 percent during that period. Natural gas became viewed as an economically efficient and environmentally friendly fuel for electric generation and industrial processing when compared with other fossil fuels.[8]

The 1990 Amendment to the Clean Air Act of 1963 was the first significant update to the act since 1970, and the first major piece of legislation calling for meaningful reductions in acid rain, urban air pollution and toxic air emissions.[9] It called for a significant reduction in the amount of sulfur dioxide (SO_2) and nitrogen oxides (NOx) emitted by coal-fired power plants over the following two decades. SO_2 and NOx are the principal pollutants that cause acid rain. According to the US EPA, in 2008, power plants burning coal and heavy oil produced over two-thirds of annual SO_2 emissions and 40 percent of NOx emissions in the United States.[10]

To comply with the Clean Air Act Amendments of 1990, electric utilities had a choice: they could either switch to low-sulfur coal, add equipment (e.g., scrubbers) to existing coal-fired power plants to remove SO_2

and NOx emissions or close down their biggest polluters—the plants that would be the most expensive to retrofit with emissions-reducing equipment. Many power producers found the cost to retrofit their older and dirtier plants prohibitive, and often turned to natural gas-fired power plants to replace coal plants.

The Birth of the Independent Power Industry

One of the largest beneficiaries of the Clean Air Act Amendments of 1990 was a new class of electricity-generating company called the exempt wholesale generator (EWG). The EWG was legislated into existence with the passage of the Energy Policy Act of 1992, and would soon become a driving force in the coming upheaval of the electric power industry. This entirely new class of utility, free from many of the constraints specified in the 1935 Public Utilities Holding Company Act (PUHCA), had the ability to price electricity according to market forces. (The PUHCA was passed in an effort to limit the size and scope of public utilities and to prohibit them from operating unregulated businesses.) With regulated electricity prices rising across the country during the 1990s due to booming economic growth, the independent power producer (IPP) emerged to capture profits in the bold new world of deregulated electricity prices. Because building natural gas-fired power plants is significantly cheaper and quicker than building either coal or nuclear plants, natural gas became the feedstock of choice for the burgeoning IPP industry. Fundamental to the business plan of virtually every IPP was "spark spread," an industry term for the arbitrage between the price of natural gas for IPPs and traditional utilities and the price they receive for their electricity. The increased efficiency of natural gas turbines for new electricity-generating stations combined with low natural gas prices and rising electricity prices—especially in West Coast markets—to make spark spreads quite healthy in the 1990s.

However, a major impediment to the early growth of the IPP industry was its inability to secure financing, since banks were wary of lending to companies that could not secure long-term supplies of natural gas on favorable terms. Without a fixed price for long-term supplies of natural gas, IPPs risk being hit by negative spark spreads, should natural gas prices rise and electricity prices remain stable or fall. It should be noted that natural gas futures began trading on the NYMEX in 1990 with little fanfare and volume and the many over-the-counter (OTC) natural gas forward contracts that are very prevalent today were largely non-existent at the time. But it would not be long before a solution emerged.

Enter the now infamous Jeffrey Skilling, who, while working in 1989 as a McKinsey Consultant employed by Enron, came up with a novel concept to help power producers secure long-term gas supplies: a scheme he appropriately dubbed the "Gas Bank":

> In Skilling's model gas producers were "depositors" in the imaginary bank. Gas consumers were the "borrowers." The producers liked the idea because Enron could give them long-term contracts for their gas, and therefore, predictable cash flow, which allowed them to plan their exploration and drilling budgets over a longer term. Gas users liked it because they would be able to predict fuel costs over multiyear terms.[11]

To secure supply "deposits" for the Gas Bank, Enron entered into long-term supply agreements with producers, who often—interestingly—received financing from Enron Finance Corp. With this steady source of supply, Enron then entered into long-term supply contracts with "borrowers" (often IPPs) on favorable terms. Despite paying higher-than-market prices for natural gas, most operators of natural gas-fired power plants were able to generate handsome returns by selling electricity with generous spark spreads into regulated markets. With the pieces in place to ensure a secure source of long-term gas supplies, lenders were increasingly willing to finance new natural gas-fired power plants for the IPP industry. Though Enron was clearly a first mover in the industry, it soon found itself with plenty of competition. Companies such as Aquila Inc., Dynegy Corp. and Calpine Corp. grew very quickly in the 1990s, due to wide spark spreads and large amounts of Wall Street money. However, a combination of too much leverage, the recession of 2001–2002 and the high-profile implosion of Enron Corporation put an end to a decade of uninhibited growth in the IPP sector.

TABLE 3.02. Natural Gas Consumed for
Electricity Generation (TCF), 1990–2000.[12]

Year	Consumed for electricity generation (tcf)*
1990	3,245
1991	3,316
1992	3,448
1993	3,473
1994	3,903
1995	4,237
1996	3,807
1997	4,065
1998	4,588
1999	4,820
2000	5,206

* Includes combined-heat-and-power plants
Source: eia.doe.gov/emeu/aer/pdf/pages/sec6_13.pdf.

Due in large part to this growth, natural gas demand from the electricity-generating sector grew a whopping 60 percent in the decade between 1990 and 2000.

The Age of Volatility Begins

Growth in demand for natural gas for electricity generation boomed throughout the 1990s—but production was not keeping pace with demand growth, despite a 55 percent increase in the natural gas-directed rig count and a 26 percent increase in the number of producing natural gas wells between 1990 and 2000 (see Tables 3.01 and 3.02). Periods of stagnating supply and growing demand are a recipe for price volatility, and that is exactly what happened. The world received a prelude to the coming age of natural gas volatility in the waning weeks of 2000, when a severe cold spell gripped the Western United States and sent natural gas prices spiraling upwards. Natural gas prices went from $4.40 on November 1, 2000 to $10.49 per mcf on December 21, 2000, an incredible 138 percent rise in only 51 days.[13]

Figure 3.01 on the next page shows the sharp rise in prices.[14]

I first learned of the spike in natural gas price when I was returning from a vacation to Costa Rica. I was spending the night before my trip home at a hotel near the airport. When I logged onto an hourly Internet

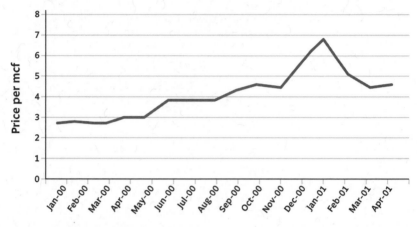

FIGURE 3.01. US Natural Gas Well Head Nominal Prices, January 2000–April 2001

computer in the hotel's lobby and saw the data, I was so shocked at the movement in price that I asked the gentlemen at the desk whether the prices the computer was showing were in local currency or dollars. I wondered if I had completely lost my mind from too much sunshine. Turns out, natural gas prices had just experienced their biggest jump in more than 16 years.

But it didn't stop there. Soon to come was nothing less than a mad scramble for gas as demand began to outstrip supplies. In the next chapter I examine how the changing natural gas market dynamics after the turn of the millennium made natural gas prices among the most volatile of any commodity over the next decade.

CHAPTER 4

2001 to 2010:
The Supply Scramble Begins

FOUR VERY IMPORTANT developments transformed America's natural gas industry in the first decade of the second millennium:

1. The peaking and subsequent decline of Canadian natural gas imports;
2. Rising LNG imports;
3. The decline of American natural gas production between 2002 and 2007, despite rising prices and increased natural gas-directed drilling activity; and
4. The emergence of shale gas as an important source of supply.

In this chapter I will also examine America's newest and most controversial source of natural gas: shale gas. While shale gas is consistently mentioned by many within the industry and the media as a major factor in the "100-year supply myth," I will provide substantial evidence that the impact shale gas will have on America's energy future is vastly overstated. Additionally I will make a realistic, though very rough, estimate of future shale gas production based on the production history of shale basins already under development. Currently, virtually all estimates of shale gas resources focus on "technically recoverable" shale resources rather than a realistic assessment of future deliverability. (I will greatly

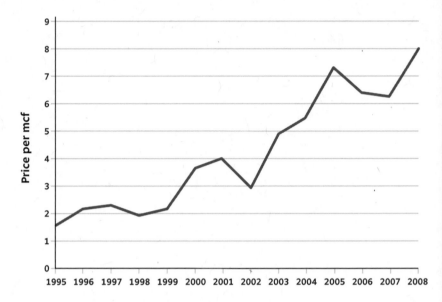

FIGURE 4.01. US Natural Gas Well Head Nominal Prices.[1]

expand my discussion of shale gas when I examine each of America's major fields in Part III.)

While US domestic marketed natural gas production grew only modestly between 1990 and 2002, imports from Canada boomed. As evidenced by Table 4.01, this period witnessed an ever-increasing torrent of natural gas pouring in from Canada. In fact, our imports increased a staggering 161 percent in only 12 years—while US domestic production grew only 6.9 percent.

By 2002, the year that marked the first important peak for net natural gas imports from Canada, Canadian imports accounted for approximately 16 percent of total US gas supplies. Natural gas production peaked in Canada in 2002, due to the advancing maturity of many of the country's largest fields and, sensibly, their exports to the US peaked the same year. Canada's production decline was not for a lack of trying. The country's average natural gas-directed rig count increased from 179 rigs running in 1999 to 271 rigs in 2003.[6] It was not until the tail end of Canada's drilling frenzy—in 2007, a year after the average annual rig count reached 361—that exports to the US would match their all-time peak 2002 levels...before falling off for a *second* time. Despite the dis-

TABLE 4.01. US NG Imports from Canada and US Marketed Production.[2,3,4,5]

Year	Canadian natural gas production (tcf)	Canadian natural gas rig count	Net US imports from canada (tcf)	US marketed production (tcf)
1990	3.49	NA	1.45	18.59
1991	3.72	NA	1.71	18.53
1992	4.11	NA	2.09	18.71
1993	4.55	NA	2.27	18.98
1994	4.90	NA	2.57	19.71
1995	5.23	NA	2.82	19.51
1996	5.42	NA	2.88	19.81
1997	5.51	NA	2.90	19.87
1998	5.66	NA	3.05	19.96
1999	5.73	179	3.37	19.80
2000	5.92	209	3.54	20.20
2001	6.05	231	3.73	20.57
2002	6.08	184	3.78	19.88

Source: EIA, Statistics Canada, Baker Hughes.

TABLE 4.02. Canadian NG Production, Canadian NG Count and US Net NG Imports from Canada.[8,9,10]

Year	Canadian natural gas production (tcf)	Canadian natural gas rig count	Net US imports from canada (tcf)
2003	5.88	271	3.43
2004	5.91	275	3.60
2005	6.03	355	3.70
2006	6.06	361	3.58
2007	5.83	215	3.78
2008	5.62	220	3.59
2009	5.21	122	3.27
2010	5.10	148	3.28

Source: EIA, Statistics Canada, Baker Hughes.

covery of two of North America's most prospective shale plays, Canada's production has not been able to eclipse 2002 levels. In fact, as of late 2012, Canadian production has fallen approximately 25 percent from its peak and continues to decline.[7]

Even with this drop in supply from our northern neighbor, US consumers of natural gas would have had little reason for concern—*if* the US had been able to increase its own production. However, marketed US natural gas production reached a post-1973 peak in 2001 of 56.35 bcf/d.[11] *After the 2001 peak, it continued to fall for the next four years, despite rising prices, a continuous increase in the US natural gas-directed rig count and a record number of natural gas wells on production.* As you can see from Table 4.03, US marketed production would not surpass its 2001 level until 2008. It should be noted that between 2000 and 2008, a ramp-up in CBM drilling in both the US and Canada somewhat reduced per well productivity, since CBM wells are far less productive than wells on the Gulf Coast, for example. However, even when considering the changed make-up of US drilling activity, producers were drilling more wells and producing less gas.

Though the modern shale gas industry was in the process of a large ramp-up in production between 2002 and 2008, albeit from a modest base, production from shale reservoirs was not enough to offset a slowdown in growth from coal bed methane (CBM) and falling production from both conventional onshore fields and the Gulf of Mexico (GOM). Marketed production from US GOM federal waters peaked in 1997 at 14.26 bcf/d and fell to an average of 4.98 bcf/d in 2011, a 65-percent drop,[16] and continues to fall. Furthermore, though hurricanes Katrina and Rita took an estimated 803.6 bcf of gas production offline between August 26, 2005 and June 19, 2006, US marketed natural gas production was not even holding steady prior to Katrina and Rita.[17] In fact, US marketed *production dropped nearly 2 percent from 2002 to 2004, despite a 48-percent increase in the natural gas-directed rig count, and an incredible 85-percent increase in average wellhead prices.*

TABLE 4.03. US Natural Gas Production, Prices, NG Rig Count, Number of Producing Wells 2000–2008.[12,13,14,15]

Year	US marketed NG production (tcf)	Average US NG rig count	Average US NG price per mcf	Number of gas producing wells
2000	20.20	720	$3.68	341,678
2001	20.57	939	$4.00	373,304
2002	19.88	691	$2.95	387,772
2003	19.97	872	$4.88	393,327
2004	19.52	1,025	$5.46	406,147
2005	18.93	1,186	$7.33	425,887
2006	19.41	1,372	$6.39	440,516
2007	20.20	1,466	$6.25	452,945
2008	21.11	1,491	$7.97	476,652

Source: EIA, Baker Hughes Incorporated.

America Turns to LNG in Response to Falling Supply

Industry's response to the America's second natural gas production peak was to look to the world LNG market for additional sources of supply. By 2003, the country's three operating LNG import terminals were in expansion mode and plans were in place to recommission a fourth terminal that had been built—and mothballed—in the 1970s. Additionally, between 2001 and 2006 more than two dozen import terminals were proposed by various companies in order to tap into the burgeoning worldwide LNG market.[18] There was strong local opposition in the potential site areas, partly because people feared that LNG terminals would be potential terrorist targets. Despite the hurdles that faced the build-out of LNG import capacity, eight new terminals were built and all four of America's existing terminals were expanded.[19] Current LNG import capacity stands at 19 bcf/d, with no new facilities likely to be built in the US for the foreseeable future.[20] Four LNG *export* facilities have been proposed to FERC in recent years, one of which is already under construction,[21] due to the pricing arbitrage between US and overseas natural gas prices.

Though US LNG imports topped out in 2007 at 2.1 bcf/d or approximately 3.5 percent of total US natural gas demand, they certainly helped to prevent natural gas prices from spiraling out of control—and inflicting substantial damage to the economy—during the first decade of the twenty-first century. As evidenced by Table 4.04, US LNG import terminal operators aggressively ramped up activity from 2001 to 2007 to take advantage of high domestic prices.

After rising 224 percent between 2001 and 2007, US LNG imports were unable to continue to increase further, largely due to the inability of import terminal operators to secure additional LNG cargoes on the spot market. In fact, by 2007, due to a major increase in LNG imports from Asian and European countries, worldwide demand for LNG cargoes began to outstrip supply. The limited number of cargoes available on the world spot market caused US imports to fall 54 percent, to 0.96 bcf/d in 2008, despite record prices for natural gas.

I consider increased reliance on the world LNG market to be extremely bad policy, for a number of reasons. First, while additional LNG

TABLE 4.04. US LNG Import 2001–2007.[22,23,24]

Year	LNG imports (bcf)	US natural gas wellhead prices	LNG imports as percent of total US consumption
2001	238	$4.00	1.07%
2002	229	$2.95	0.99%
2003	507	$4.88	2.27%
2004	652	$5.46	2.91%
2005	631	$7.33	2.87%
2006	584	$6.39	2.69%
2007	771	$6.25	3.33%
2008	352	$7.97	1.51%
2009	452	$3.67	1.98%
2010	431	$4.16	1.79%

Source: EIA.

liquefaction capacity is scheduled to come online over the next several years (largely from Australia), surging import demand from Asia and the Middle East will likely keep the world LNG market tight for the foreseeable future. In Asia, according to a February 2010 report by investment bank UBS, China is aggressively building out its LNG import capacity, which stood at a mere 0.5 bcf/d in 2008.[25] The report predicts that by 2015 China will have re-gasification capacity of approximately 5.9 bcf/d.[26] In other words, in the span of less than seven years, China is likely to have expanded its LNG import capacity eleven-fold. After a horrific nuclear disaster in March 2011, Japan, already the largest importer of LNG in the world, significantly increased its imports to make up for the loss of nuclear generating capacity.

The maturation of many of the Middle East's largest oil producing fields is causing a sea change in how the region views natural gas. For much of the history of the oil industry in the Middle East, there was no market for natural gas and producers flared trillions of cubic feet into the atmosphere. Due to the low cost of producing oil in the Middle East, nearly all of the countries in the region built their economies around this cheap source of energy. However, the combination of booming populations and increased costs to produce oil has now caused countries such as Kuwait to convert their desalination plants and electricity generation facilities from running on oil to being fueled by natural gas. The government of Kuwait has publicly stated that it wants to reduce the country's internal consumption of oil by 100,000 barrels per day and increase its consumption of natural gas in an effort to increase the country's oil exports.[27] With Brent oil prices over $110 per barrel in mid-2012 and world LNG prices at a discount to oil on a BTU equivalent basis, there is a strong incentive to replace internal oil consumption with natural gas and increase oil exports. And since Kuwait cannot meet its increased internal demand for natural gas, it now *imports* LNG from Qatar. Dubai, a country long on unoccupied buildings and short on oil, also imports LNG from Qatar. Though not an importer of LNG, Iran, owner of the enormous South Pars offshore gas field, now imports gas via pipeline from Turkmenistan since it cannot meet internal demand with domestic production. While Dubai, Iran and Kuwait are currently relatively small importers of natural gas, they are part of a much larger movement underway in the Middle East: to consume increasing amounts of natural gas internally to free up oil for export.

Won't the enormous build-out of LNG export capacity in countries such as Qatar and Australia keep the world supplied with ample amounts of LNG for decades? No way. Let's examine the LNG industry in Qatar, home to the world's largest natural gas field, the North Field. (I will discuss Australia's LNG expansion plans in Chapter 18.) Qatar, which accounted for 31 percent of total world LNG exports in 2011, is now the world's largest exporter of LNG.[28] The country has increased its exports from approximately 500 bcf in 2000 to 3.62 tcf in 2011 and is now exporting close to its capacity of 10.3 bcf/d.[29] More importantly, Qatar is unlikely to increase its LNG export capacity beyond 10.3 bcf/d for two reasons: the country has had a moratorium on new project development since 2005 (more on this below) and has built significant new industrial demand.[30] For example, Qatar has committed substantial amounts of natural gas to the world's largest gas-to-liquids plant (the Pearl Project), which was constructed by Shell, and has built an enormous chemical firm (Q-Chem) that is minority owned by Chevron Phillips Chemical.

Though Qatar is widely recognized to be home to one of the largest deposits of natural gas in the world—the EIA estimated the country had 896 tcf of proven reserves as of January 1, 2011—serious questions have arisen over the accuracy of the country's natural gas reserves.[31] According the EIA, the vast majority of these reserves are located in the country's massive offshore North Field. Figure 4.02 shows the field.[32]

Much of the skepticism about the size of Qatar's reserves surfaced after ConocoPhillips drilled a dry hole into the North Field in 2005 and the government subsequently halted new developments until at least 2014 in order to study the reservoir. An article by Dave Cohen titled "Questions About the World's Largest Natural Gas Field" chronicling the numerous uncertainties about the North Field's true reserves appeared on *The Oil Drum* website (www.theoildrum.com) in June 2006. In it, Cohen attributed the following quote to the late Matt Simmons, author of *Twilight in the Desert*:

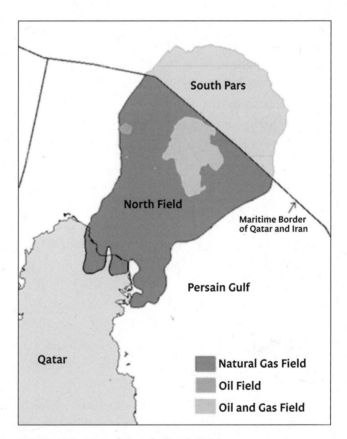

FIGURE 4.02. Map of Qatar's North Field.

Well, in 2004, if the reports were correct—and I haven't seen any denials—ExxonMobil booked 94% of its reported proven reserve additions as a result of contracts they signed in Qatar for gas from the North Field. Now, the North Field has basically two producing platforms, Alpha and Bravo. And, while ConocoPhillips last summer was drilling the wells for the Charlie platform, they hit dry holes. What's more, the quality of the gas is already sufficiently different, between Alpha and Bravo, that it would appear that the geology of the whole North Field is compartmentalized. In any event, the sheer audacity of the idea that you could have only two producing platforms in such a huge area, and know enough to book 30 years of supply is breathtaking. And we are not talking about some tiny wildcatter here. We are talking about the largest, and theoretically the most conservative, of all the oil companies in the world.[33]

There is no better example of the pitfalls of over-reliance on imported natural gas than the United Kingdom. The U.K., which imported 38 percent of its gas supply during 2010 and will increase its reliance on imports due to declining domestic production from its aging North Sea fields, provides significant insight as to what the future holds for the US as production declines accelerate and increased imports become necessary.[34] To make up for declining domestic production and increased demand, the U.K. has built Europe's largest fleet of LNG import terminals as well as the world's largest undersea natural gas pipeline, connected to an offshore Norwegian field. During the early 1980s, Margaret Thatcher's "Dash for Gas" initiative to exploit new North Sea natural gas deposits led to the building of dozens of natural-gas-fired power plants and decreased the country's reliance on coal for electricity generation. Between 1980 and 2010 the U.K. built 53 gas-fired power plants and in 2010 relied on natural gas for 40 percent of its electricity.[35] The Dash for Gas was so successful that it enabled the country to become a net exporter of natural gas to Continental Europe between 1997 and 2003.[36] However, the U.K.'s natural gas supply situation is far different today: production has declined 58 percent, from 3.828 tcf in 2000 to only

1.596 tcf in 2011, and the country now relies on imported gas to make up for falling production.[37]

The most dramatic example of the problems associated with over-reliance on imported natural gas occurred in January 2010, when a period of bitter cold descended upon much of Europe, causing a spike in demand. At the same time, the country's main import pipeline—from Norway—suffered an outage. With natural gas supplies dwindling rapidly, the U.K.'s national pipeline system operator, National Grid, was forced to issue a series of gas-balancing alerts that reduced supplies to industrial customers.[38] The alerts came as a surprise to many U.K. energy experts, as only months earlier, with storage filled to the brim and industrial demand restrained due to the recession, the country was enjoying some of the lowest gas prices in years. Though U.K. gas prices receded as the weather warmed, the country's series of gas-balancing alerts should serve as a warning sign to the country's policy makers (and to policy makers in the US as well) that over-reliance on imported natural gas is a strategy fraught with risk.

Shale Gas to the Rescue

While US imports of LNG have not met the expectations of many, both in the industry and in Washington, the commercialization of the Barnett Shale in the Fort Worth Basin of Texas in early 2004 gave renewed hope to believers in natural gas—and in America's energy future—that all was not lost. Shale gas, natural gas trapped in shale formations, quickly became the shining star on America's energy horizon by the mid-2000s.[39] Shale gas has been produced in the United States since 1821, when William Hart dug the first well specifically designed to produce natural gas. Hart's 27-foot well, which was situated along the banks of the Canaday Creek in Fredonia, New York, tapped into the state's Devonian Shale.[40] Unlike the Barnett Shale of the Fort Worth Basin, which requires man-made fractures to allow gas to flow to a wellbore, the Devonian Shale of western New York is endowed with a matrix of natural fractures that allow gas to flow to a wellbore on its own. There were also several other areas in Appalachia where early producers were able to tap into naturally fractured shale. But tapping into naturally fractured shale with very rudimentary technology and producing from a well that gets its gas from man-made fractures and what I call, modern shale gas production are two very different things. To elaborate: I define modern shale gas production as gas production from shale reservoirs requiring a man-made fracture system due to their low permeability—accomplished through hydraulic fracturing.

In 2000, after years of trial and error, the engineers at Mitchell Energy discovered that through hydraulic fracturing (pumping sand and water into man-made cracks that were created during the perforation of a well), the Barnett Shale would produce commercial quantities of gas. They had long known of the presence of natural gas in the Barnett but had not been able to tap into it until this breakthrough. News of Mitchell Energy's success spread quickly and a land rush was on. (Mitchell Energy was purchased in 2002 by Devon Energy for $3.5 billion, due in large part to their leading Barnett Shale position.)[41] Eventually, more than two dozen companies staked out acreage in Johnson, Tarrant and Wise counties and the surrounding areas, as drilling results from the Barnett continued to improve.[42] By the end of 2008, the Barnett Shale was the largest producing gas field in the United States, with an average daily production rate of nearly 5 bcf/d.[43]

By the mid-2000s, with natural gas prices continuing to rise and US conventional gas production continuing to fall—despite increased drilling—hundreds of operators began a frantic search for the next Barnett Shale. Using many of the same geological mapping techniques that led to the identification and commercialization of the Barnett, Southwestern Energy (NYSE:SWN) pioneered the commercial development of the Fayetteville Shale in central Arkansas during 2005. While not as large as the Barnett, the Fayetteville has since grown into a significant producer, with production hitting approximately 2 bcf/d by 2011.[44] More importantly, success there showed the oil and gas industry that the Barnett was not a fluke and could be repeated—*should the right geological conditions be present.*

In addition to the successes in the Barnett and Fayetteville, four other shale gas plays (Haynesville, Marcellus, Woodford and Eagle Ford) are currently under commercial development. By the end of 2009,

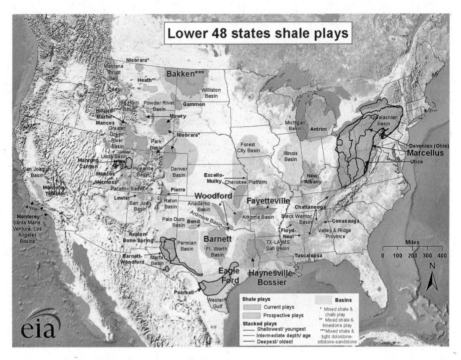

FIGURE 4.03. Shale Gas Plays, Lower 48 States.[45]

production from these six plays, along with Michigan's Antrim Shale (a shallow Devonian-age shale that was mostly developed during the 1990s and early 2000s), had reached approximately 14 percent (8.5 bcf/d) of total US natural gas production[46]; by mid-2012 it had reached approximately 19 bcf/d, or nearly one-third of total US production. With the increased usage of horizontal drilling and multi-stage hydraulic fracturing—which significantly increased initial production rates as well as costs—it seemed as though the US had finally found the answer to a secure energy future. Or had it?

By early 2010, the limits of shale gas production growth began to appear. Largely due to the very high production decline rates shale wells experience in their first two years of production, production from the Barnett Shale began to level off. After peaking at slightly more than 5.45 bcf/d in December 2010, production from the Barnett has remained fairly stable, due largely to increased in-fill drilling and the completion and tie-in of wells drilled in 2008 and left standing due to the collapse in gas prices.[47] A closer look at the Barnett Shale will be helpful in understanding some of the myths surrounding the potential of shale gas, since it is the modern shale play with the most significant production history. As of mid-2012, more than sixteen thousand wells had been drilled into the Barnett Shale and the play had produced more than 10 tcf of gas.[48] To put this in perspective, the shale play with the second longest production history in the US is the Fayetteville, which, by mid-2012, had produced approximately 3 tcf from more than four thousand wells.[49] Due to the substantial amount of data available for the Barnett, we can critically examine three commonly held myths that are the basis for the unrealistic expectations for shale gas.

Myth #1: The Manufacturing Myth

In the early years of the Barnett Shale's development, many companies leased acreage far outside the now-established core area in Johnson, Tarrant, Denton and Wise counties, in the hopes that commercially economic shale would extend beyond the known boundary of the field. And though there have been instances where operators have drilled economic wells outside of the core area, these have been the exception rather than the rule. The wide-scale leasing of acreage on the fringes of the play was due in large part to the mistaken belief that all Barnett Shale acreage was homogenous. Many promoters of early shale gas development told the investment community and the media that to understand the potential of shale gas, a new line of thinking was needed. They suggested that a "manufacturing model" approach be employed since, according to the promoters, all shale acreage had similar potential. However, the history of the more than sixteen thousand wells that have been drilled into the Barnett clearly paints a very different picture.

No one has done more to advance the understanding that shale plays are heterogeneous and that the manufacturing model is invalid than geologist Arthur Berman. Berman, who heads up geological consulting firm Labyrinth Consulting Services, suggests that a shale play's "sweet spot"—the part that contains its best wells—is due in part to well-defined structural traps, which greatly enhance productivity. In a excellent article that appeared in *World Oil* on February 22, 2010, titled

"ExxonMobil's Acquisition of XTO Energy: The Fallacy of the Manu-facturing Model in Shale Plays," he had the following to say about the idea that all shale gas acreage is created equal:

> Operators represent shale plays as low- to no-risk ventures in which gas is ubiquitous, and success can be achieved and re-peated through horizontal drilling and fracture stimulation. They have developed a manufacturing model for these plays in which the fundamental elements of petroleum geology—trap, reser-voir, charge and seal—are not critical. This appealing model has not been supported by production results. ExxonMobil probably sees a competitive advantage in taking a different approach than competitors who embrace the manufacturing model....
>
> The manufacturing model developed in the Barnett Shale play (Fort Worth Basin, Texas), where almost 14,000 wells have been drilled. The greatest number of commercially successful wells are located in two core areas or "sweet spots," and results are not uniform or repeatable even within these core areas. The Barnett Shale play is largely non-commercial because the con-trols on production are complex and difficult to predict.[50]

As can be seen in Figures 4.04 and 4.05, the core areas of the Barnett Shale are squarely located where the Muenster Arch and the Ouachita Thrust Front meet. The convergence of these two thrust belts helps to create the geological requirements for shale gas production.

Even Aubrey McClendon, CEO of Chesapeake Energy, America's second largest natural gas producer and most aggressive shale gas driller, admitted on Bloomberg News on October 14, 2009, that there are wide variations in the quality of Barnett Shale acreage when he made the fol-lowing comment:

> There was a time you all were told that any of the 17 counties in the Barnett Shale play would be as good as any other county.... We found out that there are about two and a half counties where you really want to be.[53]

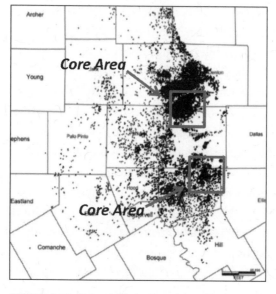

FIGURE 4.04. Barnett Shale Core Areas.[51]

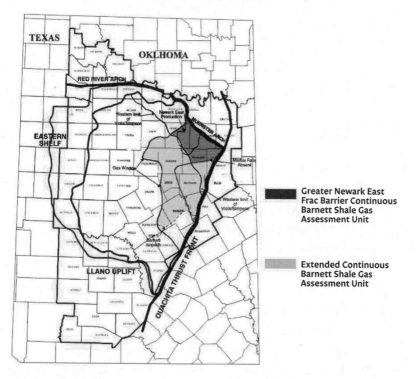

FIGURE 4.05. Important Geological Features of the Barnett Shale.[52]

It is difficult to overstate the importance of the manufacturing model to a recent significant change in the way shale gas reserves are booked. Faced with the possibility of huge reserve write-downs due to the drop-off in gas prices after the 2008 financial crisis and many shale wells not living up to expectations, the shale industry lobbied the Security Exchange Commission (S.E.C.) aggressively in 2008 and 2009 to allow for the booking of more proven undeveloped drilling locations and reserves. Previously, companies were only able to book proven undeveloped reserves associated with one undrilled location that was directly offsetting a producing well. The S.E.C. acquiesced, and beginning in 2010, allowed for the booking of multiple proven undeveloped drilling locations and their associated reserves that were not directly offsetting existing production on the condition that the wells be drilled within five years as part of its "Modernization of Oil and Gas Reporting Requirements." The justification for the rule change was the mistaken belief that shale reservoirs are homogeneous and drilling results are predictable. As we have seen this is clearly not the case. The rule change allowed some shale gas operators to not only avoid taking a write-down on their proven reserves but to show significant *growth* in proven reserves, thus keeping their all-important finding and development (F&D) costs artificially low.

Myth #2: The Decline Curve Myth

Many shale gas operators use overly optimistic assumptions, such as including the transient flow (the first months of flush production) in a decline curve analysis, and/or overly optimistic terminal decline rates to come up with estimated ultimate recoveries (EURs) that are double what a typical well could realistically be expected to produce. Transient flow can last anywhere from a few months to two years depending on the characteristics of the reservoir a well is tapping into. Once again, citing the work of Arthur Berman, who has examined the actual performance for thousands of wells of various vintages in the Barnett Shale, actual results seem to differ substantially from projections. Table 4.05 was taken from a presentation Berman gave in April 2011 in which he projected EURs from nearly 2,300 wells drilled by Devon Energy, the largest operator in the Barnett.

TABLE 4.05. Devon Energy Barnett Shale Estimated Ultimate Recovery Examples.[54]

Index	Well vintage	EUR	Number of wells
1	DVN 2004	1,555,508	91
2	DVN 2005	1,172,450	160
3	DVN 2006	1,034,726	289
4	DVN 2007	1,139,035	459
5	DVN 2008	1,358,715	573
6	DVN 2009	1,156,912	287
7	DVN 2010	1,074,705	431
	DVN WTD Avg	1,189,856	2290

Devon Energy claims that its average well will produce 2.2 bcf of natural gas as well as 100,000 barrels of natural gas liquids (NGLs; the component of natural gas that is a liquid at the surface in field facilities or gas-processing plants) over its life, while Berman's analysis cuts that number approximately in half.[55] Why the huge gap in expectations? It likely rests on Devon's claim that the proper way to model production from a Barnett Shale well is to use a hyperbolic rather than an exponential decline curve. What's the difference and why does it matter? An exponential curve is one where a well shows a steady decline throughout its life, while a hyperbolic decline curve experiences an initial period of fairly steep decline before flattening. Between three and five years of data is often required to determine whether a reservoir will have wells with an exponential decline or a hyperbolic decline and even with significant production data, results are open to interpretation.

Figure 4.06 shows how a Barnett well decline curve can contain both a transient portion and an exponential portion.

Figure 4.07 provides examples of exponential, harmonic and hyperbolic decline curves and their accompanying b-factors.

While an in-depth discussion of decline curve analysis is beyond the scope of this book, a short review of b-factor, a formula used by petroleum engineers to model a well's decline curve, is useful. In brief, the higher the b-factor, the shallower the decline curve and the longer the productive life of the well. It is considered good practice in petroleum engineering to exclude a well's transient period from a decline curve analysis since a valid decline curve analysis cannot be done until a well has reached a steady state decline. Research done by Michael

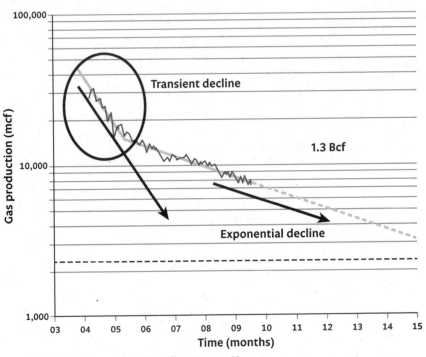

FIGURE 4.06. Barnett Well Decline Curves.[56]

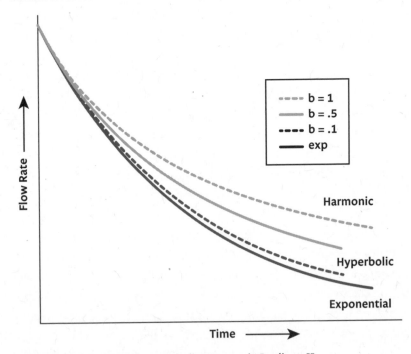

FIGURE 4.07. Exponential, Hyperbolic, Harmonic Declines.[57]

J. Fetkovitch et al., one of the nation's leading experts on decline curve analysis, described in a peer-reviewed Society of Petroleum Engineers paper (SPE 28628), supports the exclusion of transient flow from decline curve analysis.[58] However, shale gas operators often touted b-factors of higher than 1 by including the transient portion of a shale well's life in decline curve analysis when b-factors can range between 1 and 4. One company that consistently provided overly optimistic b-factors to investors was Chesapeake Energy. Figure 4.08 was re-created from a slide that appeared in the company's 2010 Analyst and Investor Day presentation.[59]

As you can see from Figure 4.08, the company presented to investors that its four large shale gas plays at the time, the Haynesville, Marcellus, Barnett and Fayetteville, all had b-factors ranging between 1.4 and 1.6. It should be noted that by mid-2011, once it became clear the b-factors that

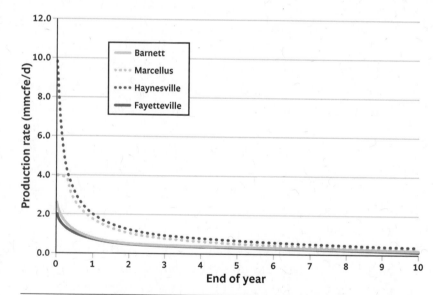

Pro forma	IP rate (mmcfe/d)	First month average production (mmcfe/d)	B-factor	Initial decline rate	Minimum decline rate	EUR (bcfe)
Haynesville	14.1	10.6	1.4	85%	5%	6.6
Marcellus	4.1	4.0	1.4	63%	5%	5.25
Barnett	3.1	2.7	1.6	74%	5%	3.0
Fayetteville	2.3	2.1	1.5	68%	5%	2.6

FIGURE 4.08. Natural Gas Shale Plays—Type Curves.

Chesapeake was presenting were quite high, all mention of shale gas play b-factors had been removed from the company's investor presentations. Also in 2011, any mention of b-factor disappeared from investor presentations of other shale gas developers.

By overstating a well's b-factor, shale operators grossly overestimate its estimated ultimate recovery and thus its economic potential. The steady state decline portion of a shale well's life will often have a b-factor of between 0.15 and 0.35 and is reflective of a well's true b-factor. Additionally, it is very rare for any well to have a b-factor of over 0.75, and those that do typically have very high permeability (a measurement of the ability of hydrocarbons to flow to a wellbore) and do not need to be hydraulically fractured. High b-factor wells can produce for decades with very shallow terminal decline curves. Modern shale wells, which require hydraulic fracturing to produce gas, typically tap into shale rock with very low permeability, and thus have relatively low b-factors.

The other decline curve assumption employed by many companies to produce overly optimistic EUR estimates is the use of unrealistically low terminal decline rates. Since shale wells are purported to have productive lives of between 40 and 65 years (more on this in Myth #3), small changes in terminal decline rates can make a big difference in a well's EUR. For example, it is common for shale companies to use terminal decline rates of between 3 and 7 percent; Devon Energy, for instance, uses a 6 percent terminal decline rate for its Barnett Shale wells.[60] However, according to work done by Arthur Berman, results from the Barnett indicate that a terminal decline rate of between 10 and 15 percent is more realistic.

I spoke with an officer of Devon about the divergence between the assumptions of Art Berman and the firm's own engineers as part of the research for this book. I wanted to see if the company could defend its decline curve assumptions. The company was unable to provide a satisfactory defense of its reserve bookings and even sent me a marketing piece, a frequently asked questions (FAQ) sheet designed to refute investor concerns about the Barnett Shale that contained many generalities and few verifiable facts.[61] The FAQ piece was produced in May 2010 after analyst Benjamin Dell of institutional research firm Sanford

C. Bernstein published a research report titled "The Death Throes of the Barnett Shale? Downgrading Devon to Market-Perform" on May 13, 2010.[62] Dell made a convincing case that Devon's operational results from the Barnett were declining and provided many facts to support his thesis. It should be noted that he independently came to largely the same conclusions about the Barnett as Berman. I contacted Dell about his research on both the Barnett and other shale plays to see if he could provide any updates to his previous pieces, since he is virtually the only Wall Street analyst to call into question the claims of the shale gas operators. However, Dell informed me that he "left the sell side and can't comment."[63] Could this be a case of Wall Street silencing an analyst who took a position contrary to an industry that generates enormous fees?

Myth #3: The 40 to 65 Reserve Life Myth

The final myth I will examine in this chapter is that shale gas wells have reserve lives of between 40 and 65 years. The use of a multi-decade reserve life grossly distorts a well's EUR, and allows for the booking of reserves that have little to no net present value (NPV). The use of a multi-decade reserve life for shale wells is one of the industry's best-kept dirty secrets. Reserve life is almost never discussed in any shale company's promotional literature and is kept out of public view as much as possible. Prior to mid-2009, I assumed that modern shale wells, given their high decline rates, low permeability and relatively short production histories, would have reserve lives of no more than 15 years. However, the Chesapeake Energy second quarter 2009 earnings conference call greatly changed my view on the length of time over which shale companies have booked their reserves. When an analyst on the call asked a question about the period over which shale reserves are booked, CEO Aubrey McClendon had the following to say:

> Yes that is 65 years. I believe that is our standard across all shale plays, which is actually a pretty interesting point to talk about. I have seen a number of other EURs from companies that are at 40 or 45, or 50 years and that actually means our curves are more conservative...[64]

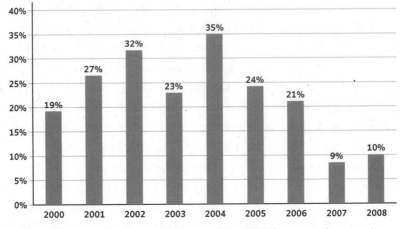

FIGURE 4.09. Barnett Wells Producing Less than 1 MMcf per Month or Dry.[65]

Like the other shale gas myths, a reserve life of more than 40 years is not supported by the data coming out of the Barnett. Once again, borrowing from the work of Arthur Berman, who has extensively studied the expected productive life of Barnett wells, we see that a significant portion of the wells that have been drilled in the past 10 years have either been plugged and abandoned or produce less than 1 million cubic feet (mmcf) per month (which generally means they are not generating enough cash flow to cover operating costs). Figure 4.09 shows the percentage of Barnett wells that have either stopped producing or produce below the 1 mmcf per month threshold.

To be fair, there is nothing illegal about using a 40 to 65 year reserve life for the booking of shale gas reserves. The US Securities and Exchange Commission allows it, though I cannot understand why, since reserves booked beyond year 12 have hardly any NPV and there is little observed evidence to suggest that most modern shale gas wells will produce for multiple decades. Given the many variables and uncertainties involved in predicting well performance over even a couple of years, the allowance of reserve bookings over multiple decades is hard to justify and grossly distorts the economics of shale gas production.

Defenders of this practice use two key points to defend it: 1) that some natural gas wells, including some vertically drilled wells producing Devonian Shale in Appalachia, have been producing for 100 years,[66] and 2) that outside reservoir engineering firms have approved the practice. I find fault with both of these arguments. As discussed earlier, the first natural gas well in the US, in Fredonia, New York, was drilled in 1821 into the shallow Devonian Shale a mere 27 feet below the surface. This well and others like it supplied the town with gas for years using rudimentary gas drilling technology. Gas simply flowed to the wellbore through large natural fractures found in the shale. However, most modern shale wells produce from rock that has very low permeability and no naturally occurring fracture system. Production history from modern shale wells suggests that without a naturally occurring fracture system, a well's productive life is likely to be far shorter than four decades.

Secondly, a third-party reservoir-engineering firm's estimate of a company's reserves, shale or otherwise, should be seen for what it is: an estimate made by humans relying on limited data and models with numerous variables. More importantly, these firms are paid by the company whose reserves they are auditing and can be subject to the same pressures financial auditing firms face. Additionally, since third-party reserve reviews are not required by the SEC, some companies only have a portion of their reserves reviewed by outside analysts. More commonly they find a third-party engineering firm who "gets" (i.e., will look favorably upon) a certain play that is important to the company and have other assets evaluated internally or by a different firm. Chesapeake Energy employed four outside reservoir engineering firms to help evaluate 77 percent of the company's proven reserves at the end of 2011.[67]

Whether reserves, shale or otherwise, are reviewed by third-party engineering firms, we have seen dozens of write-downs in recent years. Here are just a few of many examples. In the fourth quarter of 2010, Range Resources wrote down the value of its Barnett Shale assets by $463.2 million prior to their sale.[68] In Q2 2012, BHP Billiton wrote down its Fayetteville Shale assets (which it purchased from Chesapeake Energy in February 2011 for $4.75 billion) by $2.84 billion.[69] Exco Resources (NYSE:XCO), a very active driller in the Haynesville and the Marcellus, wrote down $704 million of oil and gas assets in the first half of 2012.[70] Also in Q2 2012, Chesapeake Energy wrote down 4.6 tcf of mostly proven undeveloped reserves in the Barnett and Haynesville shales.[71] Despite the recent downward reserve revisions, I still expect many more write-downs in the next few years since many shale properties are still held on operators' books at inflated values.

While I am not suggesting third-party reservoir engineering firms are the oil and gas industry's equivalent of bond rating agencies, who gave a triple-A rating to piles of subprime loans, I am suggesting that assessing oil and gas reserves is a very challenging undertaking and that oil and gas companies undoubtedly pressure reservoir engineering firms to use aggressive assumptions in their review process that may or may not prove accurate.

The Genesis of the 100-Year Supply Myth

The most commonly cited piece of evidence that shale gas is plentiful and will provide an almost never-ending bounty of energy for America is the Potential Gas Committee's (PGC) biennial report titled *Potential Supply of Natural Gas in the United States.* This assessment of potential US gas supply is ground zero for the 100-year supply myth that is now accepted as conventional wisdom. But who is the PGC and should their estimate of America's potential natural gas supply be so widely accepted?

First, the PGC is part of the gas industry-sponsored Potential Gas Agency at the Colorado School of Mines and is largely comprised of volunteer geologists and petroleum engineers. It was founded in 1960 and began publishing biennial estimates of America's natural gas supply in 1964. The research of the PGC's early days was used by industry lobbyists as a means to convince Congress that the US had sufficient natural gas resources to ensure that prices would not spike should the industry be deregulated.[72]

Though it is almost never mentioned in the mainstream media, the PGC takes care to point out in nearly all of its literature that its studies measure America's *potential natural gas resources.* What is the difference between a potential resource and a reserve? Simply put, in order to consider gas in the ground a *reserve*, it must have a 90 percent chance of being produced using existing technology, and it must produce a 10 percent rate of return using current and future natural gas prices. On the other hand, *a potential natural gas resource* is natural gas that is technically recoverable, with no consideration given either to the price at which it will be produced or when it will come out of the ground.[73] Therefore, potential natural gas resources are much less likely to ever make it out of the ground than reserves and they often have little or no production history.

An example of the huge gap between the PGC's potential resource estimate for a play and the play's probable actual production is the Marcellus Shale of Appalachia. In its December 31, 2010 assessment of America's natural gas potential, the PGC estimated that this play contains 227.3 tcf of natural gas resources, *even though at the time of this assessment the play had produced less than 1 tcf*.[74] As only 2,000 wells had been drilled into the Marcellus at the time, the committee had to rely heavily on geological models to come up with its optimistic estimate of the play's potential. More importantly, the PGC arrived at its estimate of potential gas resources in the Marcellus—and every other of the 90 gas producing basins it studies on a biennial basis—through the data and projections submitted by industry professionals working in the basins. While the PGC has taken steps to ensure that industry is not inflating the numbers it submits to them, any assumption employing limited production data should be viewed very skeptically. Based on the evolution of the Barnett Shale—which has seen its prospective area shrink massively once a significant amount of drilling occurred—I find it rather optimistic to forecast that the Marcellus contains 227.3 tcf of deliverable gas.

It is not just the PGC that faces substantial challenges when estimating hydrocarbon resources using limited data points. In 2002, the United States Geological Survey (USGS) estimated—without doing any exploratory drilling—that the National Petroleum Reserve-Alaska (NPRA) contained 10.6 billion barrels of oil and 61 tcf of natural gas.[75] However, in 2010, after analyzing 3D seismic data and drilling 30 exploratory wells, the USGS significantly reduced their previous estimate. Astoundingly, they concluded that the NPRA is likely to contain only 896 million barrels of oil and 53 tcf of natural gas, a 91.5 percent reduction in estimated oil and a 13 percent reduction in natural gas.[76]

As you can see from Figure 4.10, which was taken from the PGC's presentation announcing the results of their 2010 assessment, the

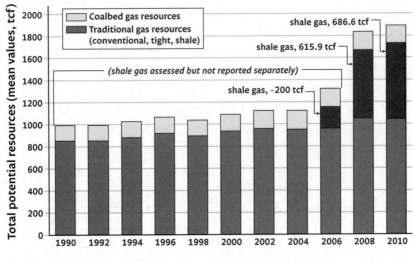

FIGURE 4.10. PGC Resource Assessment of the US, 1990–2010.[79]

committee estimates that the US has potential natural gas resources of 1,897.8 tcf.[77] Shale gas accounts for 686.6 tcf, or 36 percent of the total. It should be noted that the PGC's estimation of potential supply is *in addition* to America's total proven reserves of 272.5 tcf, as estimated by the US Department of Energy's Energy Information Agency as of December 31, 2009.[78] When combined, reserves and resources total 2,170.3 tcf of natural gas potential, or about 86 years at current consumption rates.

It is vital to remember that there is a big difference between potential resources and actual natural gas deliverability. *In other words, while the PGC rightly points out that the US has substantial natural gas in the ground, there is evidence to suggest that production may decline even at a time of rising resource estimates.* In its 2000 assessment of US natural gas resources, the PGC increased its resource base 4.4 percent from its 1998 estimate to 1,091 tcf.[80] However, as I showed in Table 4.03, US marketed production actually reached a post-1973 peak in 2001 at 20.57 tcf and did not reach this level again until 2008, after the biggest natural gas drilling boom in American history. How could this happen? With a resource base, in 2000, of nearly 50 years of consumption, how could production fall? Production declines from existing fields outstripped producers' ability to drill enough new wells to keep production stable. Remember, the PGC makes no claim on when the gas will come out of the ground or whether resources will ever be drilled.

While an in-depth technical analysis of the PGC's estimate of America's natural gas resources is beyond the scope of this book, *it should be very clear that there is a huge difference between potential resources and deliverability.* The steep production decline rates of shale wells, which can reach up to 80 percent in the first year, and the advanced age of many of America's largest gas producing basins, have created a production treadmill that is likely to result in a natural gas deliverability crisis in the 2013–15 timeframe, irrespective of the recent growth in the PGC's estimate of US natural gas resources.

While the PGC is the most widely recognized source of the 100-year supply myth, there have been other industry-sponsored studies that have also lent support to it. For example, Navigant Consulting Ltd., the American Clean Skies Foundation and the Massachusetts Institute of Technology (MIT)—all of whom have received support from the oil and gas industry—have all published studies in recent years that have unsurprisingly included optimistic outlooks on the potential of shale gas.[81] These studies should be seen for what they truly are; not objective, independent, rigorous analyses of the potential of shale gas, but marketing pieces for the industry.

The 100-year supply myth has been touted by some of its biggest beneficiaries. The following quote is an excerpt from a televised Exxon-Mobil (NYSE:XOM) ad during which one of the firm's geologists provides the company line on the potential of shale gas, even though shale is not directly mentioned in the ad:

A couple of decades ago we did not even realize how much gas was trapped in rock thousands of feet below us. Technology has made it possible to unlock this natural gas. *These deposits could provide us with enough fuel for 100 years,* providing energy security and economic growth all across this country. [Emphasis added][82]

So why is XOM spending money to make the world aware of its new-found bounty of natural gas? For two reasons. First, XOM is desperately trying to remake its image. From being thought of as one of the world's largest oil producers, and routinely vilified by politicians every time gasoline prices rise, XOM became the largest US natural gas producer after its purchase of XTO Energy in 2010 and now wants to be viewed as a more environmentally friendly natural gas company. However, US natural gas remains a relatively small portion of the company's total production base. According to the company's 2010 annual report, in that year, US natural gas accounted for only 21.3 percent of the company's worldwide natural gas production and slightly less than 10 percent of the company's entire daily production on a barrel of equivalent basis.[83] Second, and more importantly, XOM is puffing up the importance of shale gas to keep legislators from passing laws that would limit the use of hydraulic fracturing. In fact, ExxonMobil made one of the closing conditions of the XTO purchase that no detrimental fracturing legislation be passed while the deal was pending. As you might expect, perhaps due to the company's lobbying efforts, no such legislation was passed and the deal went forward. But this flexing of corporate muscle was just the beginning of the great shale game. A whole new group of boosters and promoters was waiting in the wings, ready to take up the cause.

CHAPTER 5

Enter
the Shale Promoters

D ESPITE THE RELATIVELY short history of the modern shale gas industry, by the end of the first decade of the twenty-first century, there had emerged an entire class of shale gas promoters who believed that America should hitch its economic wagon to natural gas, simply because it was clean, plentiful and produced right here in the USA. Sound familiar? But why are a handful of well-recognized experts staking their reputations on the premise that shale gas will significantly alter America's energy future? Simple. *They are selling something.* These shale optimists will profit from the myth that America has found decades worth of cheap, clean and reliable energy in shale gas. Once again, I have no doubt that shale gas is an important source of supply for the US; however, its impact has been and continues to be grossly overstated.

As CEO of Chesapeake Energy (NYSE:CHK)—the second largest producer of natural gas in the US—Aubrey McClendon has done more to promote the development of shale gas than any other person in the oil and gas industry.

Let's begin with a review of his testimony before the US House Select Committee on Energy Independence & Global Warming on July 30, 2008, which was chaired by Congressman Edward Markey of Massachusetts. Here are a few highlights from McClendon's opening remarks to the Committee:

If there is one message I would like to effectively communicate today, it is that America is at the beginning of a great natural gas boom. This boom can largely solve our present energy crisis. The domestic gas industry through new technology has found enough natural gas right here in America to heat homes, generate electricity, make chemicals, plastics and fertilizers, and most importantly, potentially fuel millions of cars and trucks for decades to come.

This great new period of discovery in our industry has largely gone unnoticed by the media and most policymakers. Our industry has recently learned how to extract natural gas from massive rock formations called gas shales buried deep below the earth's surface.[1]

During the hearing's Q and A session, Chairman Markey asked McClendon if there would be enough natural gas to use as feedstock to increase output from America's natural gas-fired power plants. This is an excerpt from the hearing:

McClendon: "Three years ago I would have said no. Today I say yes because of the technological breakthrough that has occurred in drilling for natural gas into these shale deposits."

Markey: "So with the supply available, how long do you think it would take to make a transition like that?"

McClendon: "I am not an expert in that. *What I am expert in doing is telling you that I believe we can increase the supply of natural gas…by at least 5 percent per year.* [Emphasis added.] How the market decides to use that, whether it be for cars, whether it be for electricity, whether it be for more plastics and chemicals I can't really comment on, *but I think the surprise that I have for people today is that the technological breakthrough that we have developed in finding gas from shales changes everything about what you think about natural gas scarcity in America.*"[2] [Emphasis added.]

The following natural gas production forecast comes from the transcript of McClendon's Q and A session:

I am here to testify that from existing sources, areas of production—Texas, Oklahoma, Louisiana, Arkansas—that there are enormous new resources of natural gas that in my view will increase the supply of natural gas in this country for at least the next decade by a minimum of 5 percent per year.[3]

Similar to the AGA's proclamation in the 1970s that higher prices will "keep the gas coming," McClendon makes clear his view that technology has changed everything about natural gas supplies in America.[4] More specifically, his prediction that natural gas supplies will grow by a "minimum of 5 percent per year" for a decade is absurd. In his written testimony, McClendon correctly points out that US consumption in 2008 was approximately 63 bcf/d, with domestic production comprising 54 bcf/d and imports 9 bcf/d.[5] McClendon's contention implies US-marketed natural gas production would reach *87.96 bcf/d by 2018*, an increase of 63 percent from 2008 levels. As this book goes to print in early 2013, Lower-48 US natural gas production has plateaued at approximately 62 bcf/d and there is no indication that adding another 25 bcf/d of production is even remotely possible.[6]

McClendon increased his estimate of America's future deliverability of natural gas during a July 2011 media conference call to unveil CHK's investment in an alternative energy company and its "three-prong plan for gaining independence from OPEC oil."[7] Here is an excerpt from the conference call:

I think the thing to recognize here is simply the remarkable size of the American resources in terms of natural gas. People I think haven't properly focused on the fact that in 2009, for example, we passed Russia to be the largest natural gas producer in the world. I mean how many Americans today know that we're the OPEC of natural gas here in the US, the Saudi Arabia of natural gas in the US? I think that's a little known fact and so we need to do more and more to convince folks of that.... *And know that if the industry needs to produce 90 bcf a day rather than 60 or 100 bcf a day rather than 60 bcf a day we believe that's absolutely achievable in the years ahead and likely at a cost that may be even less than where it is today simply because we are always able to figure out ways to do things cheaper down the road in these shale plays.* [Emphasis added.][8]

McClendon is correct in saying that in 2009 the US became the largest producer of natural gas, once Russia smartly stopped investing in drilling after prices collapsed due to the 2008 financial crisis and resulting drop in demand for European exports. However, the comparison to OPEC—I will review the comparison to Saudi Arabia later—is inaccurate. The US is *not* a net exporter of natural gas. As I have discussed previously, for more than 25 years, the US has been a net importer of natural gas from Canada and overseas via LNG. For McClendon's claim that the US can grow from 60 bcf per day to 90 or 100 bcf per day to be realized, the country would need to add the production equivalent of more than five new Barnett Shales on top of the existing production base. This is simply not going to happen under any foreseeable scenario. Remember, due to the natural decline rate of roughly 20 percent per year, the US must bring on new wells capable of producing roughly 12 bcf a day every year just to keep production flat, a very difficult task given the advanced maturity of many US natural gas-producing fields and the high decline rates of shale wells.

In addition to vastly overstating America's natural gas deliverability, McClendon has also provided the media with an over-inflated view of America's endowment of natural gas. His most egregious predictions about the importance of shale gas came on CBS News' "60 Minutes" program that aired on November 14, 2010. In a segment titled "Shale-ionaires," McClendon had the following exchange with Leslie Stahl:

McClendon: "In the last few years we have discovered the equivalent of two Saudi Arabia's of oil in the form of natural gas in the United States. *Not one, but two.*" [Emphasis added]
Stahl: "We have twice as much natural gas in this country, is that what you are saying, than they have oil in Saudi Arabia?"
McClendon: *"I am trying to very clearly say exactly that."*[9] [Emphasis added]

Two Saudi Arabia's of natural gas? Really, Aubrey? To put into perspective how much gas the equivalent of two Saudi Arabia's worth of oil is, consider the following: according to the 2011 BP Statistical Review, Saudi Arabia had 264.5 billion barrels of proven reserves of oil as of the end of 2010.[10] This is the equivalent of 1,587 tcf of natural gas. The equivalent of two Saudis worth of oil would be 3,174 tcf of natural gas, or enough to supply the US at its current rate of consumption for approximately 125 years. McClendon's claim that the US has the natural gas equivalent of two Saudi Arabia's worth of oil is not supported by any empirical evidence whatsoever.

Given that there is no evidence to support McClendon's outlandish claims on the importance of shale gas, why is he so vehement about his message? In other words, what is he selling? In a brief moment of candor at the height of the 2008 financial crisis, McClendon made it clear to the world the importance of selling interests in his existing shale gas plays to latecomers. Below is an excerpt from CHK's Q3 2008 Business Update Conference Call on October 15, 2008:

We are also going to be able to be cash flow positive from this point forward. That includes a part of our business model that apparently some people still have a hard time understanding and I think there are two ways to make money in the business. One is to drill wells and just have the gas produce out over time.

But there are other ways as well and that is doing these various asset monetizations. I think when we're through with 2008 you will see that our company will have monetized somewhere between $10 billion to $12 billion of assets during the year including drilling carries and would have an indicated profit margin if you will on that of about $10 billion. *I can assure you that buying leases for X and selling them for 5X or 10X is a lot more profitable than trying to produce gas at $5 or $6 mcf.*[11] [Emphasis added.]

Between 2006 and 2011, Chesapeake Energy sold whole or partial interests in six of its shale gas plays for an incredible $17.19 billion when future development costs are included.[12] Five of Chesapeake's six transactions were done with foreign-based companies such as Total, BP, Statoil and BHP Billiton, eager to learn the business from one of its most active companies. While these large multinational companies have deep pockets and can spend significant sums to enter new areas, the US shale gas business has been a disaster for virtually all of them that bought shale gas assets from Chesapeake.

In addition to the $2.84 billion write-down BHP Billiton took on the assets it bought, BP also appears to have overpaid Chesapeake for shale gas assets.[13] The world got an insight into BP's desire to get into the shale gas business at the World Economic Forum meeting in Davos, Switzerland, in early 2010 when the now dethroned CEO of BP, Tony Hayward, gushed the following platitude about the importance of shale gas in the US:

> [It's] a complete game-changer in the US. It probably transforms the US energy outlook for the next 100 years.[14]

BP entered the US shale business in July 2008, at the height of the frenzy, by purchasing Chesapeake Energy's Arkoma Basin Woodford Shale assets for $1.75 billion.[15] The purchase included 90,000 net acres of leaseholds and production of 50 mmcf/d. Using an 80-acre spacing unit, which is generous since not all acreage will prove to be prospective, BP paid approximately $1.5 million in land costs per each potential well or $19,400 per acre.[16] Based on natural gas prices since the time of the purchase and the quality of the land they bought from Chesapeake— which is second-tier acreage compared to play leader Newfield Exploration's—there is virtually no foreseeable scenario under which BP will ever be able to recover its initial investment in the Woodford Shale. BP quickly followed up its Woodford purchase with a plunge into the Fayetteville Shale. In September 2008, BP announced that it paid $1.9 billion to Chesapeake Energy for a 25-percent stake in its 540,000 net acres in the company's Fayetteville Shale play.[17] The purchase included cash consideration of $1.1 billion and $800 million in drilling carries that were paid out by the end of 2009. Also included in BP's acquisition was a 25-percent stake in the 180 mmcf/d in Chesapeake's Fayetteville Shale production (45 mmcf/d net). In other words, for BP's 135,000 net acres in the Fayetteville, the company paid approximately $14,000 an acre or $1.1 million per 80-acre drilling location. Once again, due to BP's high price of entry into Fayetteville, it will likely be impossible for the company to recover its investment let alone generate a positive rate of return. As part of its Q2 2012 earnings release, BP announced that it was writing down its US shale gas assets by $2.1 billion.[18]

Another strong advocate of shale gas is none other than Pulitzer prize-winning author, former Harvard professor and Chairman of IHS CERA, Dan Yergin, PhD. IHS CERA is an energy industry consulting firm whose clients are mostly large oil and gas companies. Since 2009, Yergin has been pushing the phrase "shale gale" to describe the impact shale gas is having on America's energy supply[19] and has called it "simply the most significant energy innovation so far this century."[20] Yergin provided a great example of his incredibly optimistic view of the potential of shale gas in an April 2, 2011 article he wrote for *The Wall Street Journal* titled "Stepping on the Gas." Here is a quote from the article:

> Estimates of the entire natural-gas resource base, taking shale gas into account, are now as high as 2,500 trillion cubic feet, with a further 500 trillion cubic feet in Canada. That amounts to a more than 100-year supply of natural gas, which is used for everything from home heating and cooking to electric generation, industrial processes and petrochemical feedstocks.[21]

As a consultant to some of the biggest energy companies in the world (ExxonMobil is rumored to be a longtime client), it should come as no surprise that IHS CERA is the world leader in the lucrative business of providing political cover from Washington as "Big Oil" transitions itself into "Big Gas." As major international oil companies have seen their reserves of oil shrink and natural gas increase, Big Oil needs someone to over-inflate the importance of natural gas, especially domestic natural gas. Dan Yergin fits the bill perfectly. For example, as environmental concerns over hydraulic fracturing continue to threaten the multi-billion-dollar investments in shale gas plays by the likes of RoyalDutch Shell,

Chevron Corporation and ExxonMobil, the need for a credible industry advocate has become vital. Lawmakers are much more open to compromise with the industry on controversial subjects such as the impact of hydraulic fracturing when they are under the delusion that the US has a 100-year supply of shale gas. Hence, the importance of the 100-year supply dream.

So just how credible is Dan Yergin? For instance, let's examine his views on Peak Oil. While I will be the first to admit that making predictions about energy supplies and prices is inherently difficult, Yergin has repeatedly used woefully unrealistic supply projections in an effort to invalidate the very real concerns raised by those advocating that Peak Oil be taken seriously. Despite record oil prices and flattening world oil production in 2005, Yergin penned the following in a July 2005 *Washington Post* article titled "It's Not the End of the Oil Age":

We're not running out of oil. Not yet....

But it is oil that gets most of the attention. Prices around $60 a barrel, driven by high demand growth, are fueling the fear of imminent shortage—that the world is going to begin running out of oil in five or 10 years. This shortage, it is argued, will be amplified by the substantial and growing demand from two giants: China and India.

Yet this fear is not borne out by the fundamentals of supply. Our new, field-by-field analysis of production capacity, led by my colleagues Peter Jackson and Robert Esser, is quite at odds with the current view and leads to a strikingly different conclusion: *There will be a large, unprecedented buildup of oil supply in the next few years. Between 2004 and 2010, capacity to produce oil (not actual production) could grow by 16 million barrels a day—from 85 million barrels per day to 101 million barrels a day—a 20 percent increase. Such growth over the next few years would relieve the current pressure on supply and demand.*[22] [Emphasis added.]

Clearly, if such remarks about capacity are supposed to bear any relation to actual output, Yergin was way off the mark. In fact, according to the

2011 BP Statistical Review, world oil production increased only slightly from 81.485 million barrels per day in 2005 to 82.095 million barrels per day in 2010.[23] Yergin's firm, IHS CERA, has made even more outlandish statements about the world's potential capacity to produce oil. [Note: IHS CERA takes the wishy-washy approach of always emphasizing "potential capacity" when discussing the Peak Oil topic since it allows them to extend their credibility on the subject and never be proven wrong.] Below is an excerpt from a September 2006 *Businessweek* article discussing new deepwater discoveries in the Gulf of Mexico:

> Cambridge Energy Research Associates predicts world oil and natural gas liquids capacity could increase as much as 25% by 2015. Says Robert W. Esser, a director of CERA: *"Peak Oil theory is garbage as far as we're concerned."*[24] [Emphasis added.]

Yergin's and CERA's stance on Peak Oil runs completely contrary to the substantial evidence that Peak Oil is near or has already arrived (a topic for another book) and has drawn heavy criticism from experts on the subject. In fact, in 2008, the members of the Association for the Study of Peak Oil (ASPO)—a group of very well respected earth scientists, academics and analysts—became so disgusted with Yergin's and CERA's repeated unrealistic projections of future world oil supply capacity that several members offered to publicly debate Yergin on the subject of Peak Oil. Certain members of ASPO were so confident in their position that they offered to bet Dan Yergin $100,000 on the validity of "CERA's June 2007 forecast that world oil production capacity will reach 112 million barrels per day (mmb/d) by 2017, which extrapolates to 107 mmb/d of actual production, up from about 87 million barrels today."[25]

Here is a quote from the press release announcing the wager and the request for debate:

> "CERA is forecasting an addition of 20 million barrels within a decade," said Steve Andrews, co-founder of the Association for the Study of Peak Oil-USA (ASPO-USA). *"That's a vision in*

search of reality. Anything is possible on paper, but we are betting you can't do that with the drill bit."[26] [Emphasis added.]

While conventional wisdom still accepts Dan Yergin and CERA's nonsense that we are experiencing a "shale gale," I expect Yergin's wildly optimistic predictions about the impact of shale gas to be discredited within the next few years.

The biggest advocate of shale gas from the world of academia is Pennsylvania State University's Terry Engelder, PhD. Much of the notion that the Marcellus has nearly 500 tcf of natural gas resources can be traced to an article titled "Marcellus" Engelder penned for the *Fort Worth Basin Oil and Gas Journal* in August 2009. In the article, Engelder discussed very early well results from operators in the Marcellus and provided the following logic:

The Marcellus is prospective under at least 117 counties in five states of the Appalachian Basin. Using several geological parameters, I have graded each of these 117 counties according to a six-tier system. Using the power-law rate decline of Chesapeake, an EUR for the Marcellus may be calculated assuming that 70 percent of the sections in each county are accessible, that wells have an 80-acre spacing and that decline is allowed to proceed for 50 years. *This calculation yields a 50 percent probability that the Marcellus will ultimately yield 489 tcf.* [Emphasis added][27]

Engelder's estimate that the Marcellus shale may produce 489 tcf is 186 percent higher than the PGC's 2010 estimate of total Marcellus resource potential of 227.3 tcf.[28]

While there is little doubt that the Marcellus is a very significant natural gas resource, I find it hard to accept either Engelder's resource estimate—or the PGC's—for it. As part of the research for this book I examined Marcellus drilling activity and well results from the Pennsylvania Department of Environmental Protection's (DEP) website to come up with my own estimate of what the Marcellus could be expected to produce over its lifetime. After poring over all of the well data, it be-

TABLE 5.01. Pennsylvania Core
Counties and Acreage.[30]

County	Acres
NE Core Counties	
Bradford	736,429
Clearfield	734,285
Potter	691,949
Tioga	725,587
Susquehanna	526,630
Lycoming	790,304
SW Core Counties	
Fayette	505,690
Greene	368,550
Washington	548,538
Westmoreland	656,307
Total Acres	**6,284,269**

Source: US Census.

came clear that the vast majority of drilling up to that time occurred in two distinct core areas, one in the southwestern part of the state and another in the north-central portion. I focused my research efforts on Pennsylvania since West Virginia has very little Marcellus data available to the public and New York had placed a moratorium on hydraulic fracturing, thus eliminating all Marcellus drilling.

So how big are these two core areas and what percentage of all production is coming from them? Based on data from the DEP, I discovered that approximately 95 percent of all Marcellus production in the state for the period July 1, 2010 through December 31, 2010 came from only 10 counties. Table 5.01 shows core area counties and the number of acres contained in each.

Due to the heterogeneity of shale rock, there can be wide variations in the production potential of even the core areas of shale plays. Therefore, I would suggest that only one-third of all acreage in the 10 core counties, about 2.095 million acres, comprise the Marcellus Shale core area. Assuming future drilling in the Marcellus validates the core acreage, there is a possibility that the Marcellus may produce 50 tcf over its lifetime—a far cry from Dr. Engelder's 489 tcf but a huge amount of gas nonetheless. (I based this very crude estimate on 25,000 wells with an average estimated ultimate recovery of 2 bcf per well. While there will be many Marcellus Shale wells that will have EURs of over 4 bcf, there will also undoubtedly be many wells that will be dry or produce minimal amounts of gas.) To put into perspective how aggressive an estimate of even 50 tcf of potential production from the Marcellus Shale is, consider the following: the largest natural gas field in North America by historical production is the massive Hugoton field of western Kansas, Oklahoma and the Texas Panhandle. According to a 2006 study by the Kansas Geological Survey titled "Hugoton Asset Management Project," it has produced approximately 35 tcf of natural gas during its 70-year productive life.[31] In other words, *for the Marcellus to recover even 50 tcf, it would have to produce nearly 50 percent more gas than the largest field in North America has ever produced.* Remember, due to the very early stage nature of the Marcellus—which had yet to produce even 1 tcf of gas by the end of 2010—any projection, regardless of the source, should be taken with a fistful of salt.

Additionally, it should be noted that Engelder's August 2009 estimate of 489 tcf of potential for the Marcellus is quite confusing, bearing in mind that in May 2008, he coauthored an article in *The American Oil and Gas Reporter* titled "Marcellus Shale Play's Vast Resource Potential Creating Stir in Appalachia" in which he argued that the Marcellus may contain a mere 50 tcf of natural gas resources.[32]

So what changed between May 2008 and August 2009 that would cause Engelder to increase his estimate of the Marcellus Shale resources nearly tenfold? I asked him this exact question as part of the research for this book. Engelder attributed the increase to the use of a different estimation methodology and from production data on approximately 50 wells he gleaned from the investor presentations on the websites of Range Resources, Chesapeake Energy and other publicly traded operators active in the Marcellus Shale.[33] Do you think it's possible that these companies might have been highlighting only their best wells in their presentations to the investment community? Engelder also confirmed to me that he used no production data whatsoever from the state of Pennsylvania in his new and improved 489 tcf estimate, since the state had yet to begin publicly releasing it.

Try as I might, I cannot understand how such a large increase in the total resource of the Marcellus could possibly be justified, considering the extremely limited production data Engelder had to work with. It seems much more likely to me that his sudden projection jump was related to his work away from Penn State. Interestingly, in addition to his duties at the university, Dr. Engelder is a principal at Appalachian Fracture Systems, Inc., a geological consulting firm that consults for operators in the Marcellus Shale, among other activities.[34] Apparently, Dr. Engelder prefers to keep his relationship with Appalachian Fracture Systems out of the public spotlight for fears that it may diminish his credibility. (As of May 2011, Appalachian Fracture Systems appears nowhere on his Penn State online curriculum vitae.)[35] Does having a financial interest in the advancement of the Marcellus diminish an academic's credibility on the subject? In my opinion, absolutely!! After watching Charles Ferguson's brilliant, Academy Award-winning documentary "Inside Job," in which he interviews several academic economists who also happened to be consultants to mega-banks and authors of papers encouraging bank deregulation and limited regulation of over-the-counter derivatives, it is clear that academic credibility takes a backseat to the almighty dollar.

It is not just Terry Engelder at Penn State who has promoted the importance of the Marcellus in recent years. In a remarkably non-transparent "white paper" published in 2009 titled *An Emerging Giant: Prospects and Economic Impacts of Developing the Marcellus Shale Natural Gas Play*, professors Timothy Considine and Robert Watson grossly overstated the economic importance of the Marcellus to the state.[36] Among the claims they made was that Professor Engelder's estimate of 489 tcf of Marcellus resources seemed "increasingly reasonable." Below is an excerpt:

> While reserve estimates should be considered somewhat uncertain at this early stage, as each new Marcellus well is completed, estimates of recoverable reserves of at least 489 trillion cubic feet seem increasingly reasonable.[37]

The report also claimed that Marcellus development "could be generating $13.5 billion in value added and almost 175,000 jobs in 2020."[38] Lastly, it found that a recently proposed severance tax (a tax every major oil- and gas-producing state except Pennsylvania collects when oil and gas comes out of the ground) on natural gas by then-governor Ed Rendell "cannot be passed onto consumers and, therefore, drilling activity would decline by more than 30 percent and result in an estimated $880 million net loss in the present value of tax revenue between now and 2020."[39] While the professors are entitled to their own opinion, public advocacy group Responsible Drilling Alliance found their conclusions to be overly one-sided and took issue with the fact that they did not disclose that the study was funded by industry advocacy group the Marcellus Shale Gas Committee.[40] It should be noted that the report issued by professors Considine and Watson was widely used by advocates of Marcellus development within the oil and gas industry and elsewhere and became known as the "Penn State Report."[41] Apparently, William Easterling, Dean of the College of Earth and Mineral Sciences at Penn State, investigated the report and agreed with the Responsible Drilling Alliance.[42] In a letter to the alliance, he said:

In the initial version of the earlier report, we found flaws in the way that the report was written and presented to the public. First, the report did not identify the sponsor of the research, which is a clear error. As a matter of policy, all publications emanating from externally sponsored research at Penn State are required to identify the sponsor of the research. Second, the authors could and probably should have been more circumspect in connecting their findings to policy implications for Pennsylvania, and may well have crossed the line between policy analysis and policy advocacy. In particular, the prose in the section dealing with the potential effects of severance tax on drilling rates in Pennsylvania should be more scholarly and less advocacy-minded. Moreover, the authors should have made their points without being adversarial to the Governor.[43]

Penn State eventually retracted and later reissued the report under the new title "The Economic Impacts of the Marcellus Shale Natural Gas Play: An Update" with only minor changes, including crediting the Marcellus Shale Coalition for funding.[44]

Surprisingly, Professor Considine's new employer, the University of Wyoming, is either not aware of the controversy surrounding his research while at Penn State or does not share Penn State's view on full disclosure. In August 2012, I downloaded the original version of Considine and Watson's published paper, without the funding disclosure, from the University of Wyoming's Center for Energy Economics and Public Policy website.[45] I find it hard to believe that America's publicly funded academic institutions do not take a harder stance on lack of disclosure. Lending academic credibility to industry-sponsored research without disclosure is in my opinion woefully unprofessional and should not tolerated by any institution of higher learning.

The final widely recognized promoter of shale gas that I will discuss here is oil-and-gas-industry-veteran-turned-hedge-fund-manager T. Boone Pickens. Pickens has been a longtime believer in the importance of natural gas going all the way back to his days at Mesa Petroleum, when he was one of America's preeminent corporate raiders. In recent years, Pickens has been promoting natural gas, and especially shale gas, as a way to break the grip of foreign oil, create jobs in America and clean up the environment.[46]

On July 8, 2008, Pickens took his efforts to promote increased natural gas usage to a new level when he unveiled the Pickens Plan, which advocates for replacing America's use of foreign oil through increased natural gas consumption for transportation purposes and increased reliance on wind and solar for electricity generation. Pickens envisioned wind power generating 20 percent of the nation's electricity, and partially displacing natural gas-powered electricity. Part of the displaced natural gas could then be used for transportation purposes. In conjunction with the unveiling of his scheme, Pickens made public his plans to build the world's largest wind farm in the Panhandle of Texas.[47]

To support America's increased dependence on natural gas, the Pickens Plan website makes some grandiose claims about America's natural gas supplies. Similar to Yergin and McClendon, Pickens highlights the Potential Gas Committee's research as a major reason America should increasingly look to natural gas to augment our energy security. From the Pickens Plan website:

> The biennial study of the Potential Gas Committee (Colorado School of Mines) concluded that, when calculating the impact of shale deposits, new discoveries represent a "net increase in total potential resources of 39 percent" since 2006. *The New York Times* reports new natural gas studies demonstrate a "stunning increase in potential resources." The energy available from natural gas contained in these shale deposits can provide ample supplies for the next 100 years.[48]

To generate public interest in the plan, Pickens ran a series of television and print advertisements claiming it would "break the stranglehold of foreign oil."[49] But what really is the plan?

Many supporters of the Pickens' vision might be surprised to discover that his plan is largely a lobbying vehicle for the passage of US House Bill H.R. 1835 and US Senate Bill 1408, commonly referred to as the Natural Gas Vehicle Act. Here is a synopsis of H.R. 1835, written by the Congressional Research Service, a non-partisan research division of the Library of Congress:

> New Alternative Transportation to Give Americans Solutions Act of 2009—Amends the Internal Revenue Code to: (1) allow an excise tax credit through 2027 for alternative fuels and fuel mixtures involving compressed or liquefied natural gas; (2) allow an income tax credit through 2027 for alternative fuel motor vehicles powered by compressed or liquefied natural gas; (3) modify the tax credit percentage for alternative fuel vehicles fueled by natural gas or liquefied natural gas; (4) allow a new tax credit for the production of vehicles fueled by natural gas or liquefied natural gas; and (5) extend through 2027 the tax credit for alternative fuel vehicle refueling property expenditures for refueling prop-

erty relating to compressed or liquefied natural gas and allow an increased credit for such property. Requires 50 percent of all new vehicles purchased or placed in service by the US government by December 31, 2014, to be capable of operating on compressed or liquefied natural gas. Authorizes the Secretary of Energy to make grants to manufacturers of light and heavy duty natural gas vehicles for the development of engines that reduce emissions, improve performance and efficiency, and lower cost.[50]

There was even an electronic petition on the Pickens Plan website urging visitors to demand "an energy independence plan for America NOW!" from President Obama and the 111th US Congress. The petition calls H.R. 1835 and S. 1408—referred to by the website as the Natural Gas Act—"the best tool we've had in decades to reduce our dependence on foreign oil."[51]

By 2011, after promoting the Pickens Plans for three years and facing an unclear timetable for the passage of natural gas vehicle legislation, Pickens stepped up his appeals to the American people by ensuring them that there is plenty of natural gas—enough for generations. Below is an exchange he had with hosts from CNBC in April 2011, in which he provides his estimate of America's natural gas endowment and its importance to the country:

Pickens: Well you've got to start somewhere. And ya know if you sat down and looked at all the ways we haven't managed energy in America, that if you looked at it, you'd cry. So you say OK, having cried, now what are we going to do about it? You've got to get on something. You've got to make something happen. You just can't sit there and get [sic] take more and more oil from the enemy and we're taking five million barrels of oil from OPEC everyday. Take the eight million trucks go to natural gas, you could cut OPEC in half. But the thing I would like to comment about is that today, if I announced that we have more oil equivalent than the Saudis do, I would be telling you the truth.

Host: In the United States?

Pickens: In the United States.

Host: Natural gas?

Pickens: Natural gas. I mean this isn't my number. This is other people's.

Host: That isn't the way it was five year ago. Because of fracking and everything else, you're there.

Pickens: *You're there. You're there. I say you're going to recover 4,000 trillion. Which is 700 billion barrels.*[52] [Emphasis added.]

4,000 trillion cubic feet? Now that is a huge number! 4,000 tcf is roughly the equivalent of 160 years of US natural gas production at current rates and Pickens provides no evidence to support his claim during the interview or on his website. Similar to all the other shale promoters, he is long on providing attention-grabbing estimates on the potential of natural gas but short on offering details on the fields that will supply this bounty.

In an age where tremendous lip service is paid to transparency, it should come as no surprise that Pickens stands to profit handsomely should developers of natural gas vehicles receive huge government subsidies and tax incentives. According to documents filed with the US Securities and Exchange Commission, as of December 31, 2011, Pickens owned approximately 16.5 million shares (valued at over $200 million as of August 2012) of Clean Energy Fuels Corp. (NASDAQ:CLNE), one of the largest providers of natural gas refueling stations in the US. Needless to say, CLNE is uniquely positioned to profit from the Natural Gas Vehicle Act.[53]

In addition to promoting the Pickens Plan and subsidies for natural gas vehicles, Pickens has also collected funds from the American public via his website and spent millions of his own funds to further his very self-interested agenda.[54] However, very little progress has been made on delivering the promises of his Plan. As of late 2012, natural gas vehicle subsidies appear to be going nowhere in Congress and Mesa Power Group LLC—the company Pickens set up to build what he touted as the world's largest wind farm in the Texas Panhandle—has yet to generate its first watt of commercial wind power.

A Realistic Estimate of Future Shale Gas Production

So what is a reasonable estimate of America's endowment of shale gas? While we are still early in the development of the modern shale gas industry and both shale gas companies and industry-sponsored entities have put many hyperbole-filled estimates on its potential into the public domain, there has been little fact-based analysis of its production potential. Nearly all of the estimates of shale plays published to date have focused on "technically recoverable" resources, regardless of whether commercial production has been established there or not. I have chosen a more practical way of examining the importance of shale gas by using known production history as the basis of my analysis. By briefly examining the production history of both the major and minor modern shale plays, as well as a few of the plays that once held great promise and now lie fallow, we can gain insight into a reasonable estimate for future shale gas production. Future production is what will heat homes, generate electricity and make fertilizers and chemicals, while calculating technically recoverable resources is largely an academic exercise. Though technology will undoubtedly advance and open up previously out of reach shale resources, in making the below estimate, I am not counting on unpredictable technological advances to add to the contribution shale gas will make to America's future natural gas production. I am simply using existing production history—the most reliable predictor of future performance—as well as some logic, to make a very rough estimate of future production.

Major Shale Plays (Current Production > 500 mmcf/d)

1. **Barnett Shale:** Having already produced 10 tcf as of mid-2012, the Barnett is the granddaddy of all modern shale plays, and has accounted for approximately 50 percent of all modern shale gas production.[55] Given that more than 16,000 wells have been drilled and the outer limits of the play already defined, assigning the Barnett 25 *tcf* of additional future production would be a generous estimate.[56]

2. **Haynesville Shale:** Haynesville production grew rapidly between 2008 and 2011 and appears to have peaked at approximately 6 bcf/d.[57]

This level of production—much of which is woefully uneconomic at gas prices below $6 per mcf—only came after operators drilled many wells to meet lease commitments. (Oil and gas leases in the Haynesville play often require a well to be drilled within three years of signing.) The State of Louisiana does not provide Haynesville-specific production data, which makes any forecast of future production somewhat difficult. However, based on publicly available information, as of mid-2012, the play has produced roughly 5 tcf in total.[58]

In an article in the December 2011 issue of *The Oil and Gas Journal* titled "Louisiana Haynesville-1: Characteristics, Production Potential of Haynesville Shale Wells Described," Mark J. Kaiser and Yunke Yu of the Louisiana State University Center for Energy Studies stated that the average Haynesville well will produce approximately 3 bcf over its lifetime.[59] It should be noted that Kaiser and Yunke independently came to the same estimate as Art Berman and Lynn Pittinger did in their August 2011 article "US Shale Gas: Less Abundance, Higher Cost" that appeared on the theoildrum.com.[60] More importantly, similar to the projections of Berman and Pittinger, Kaiser and Yu noted that their "EUR (estimated ultimate recovery) values appear low relative to company estimates and industry expectations."[61] Due to the relatively small core area of the Haynesville—estimated to be between 100,000 to 150,000 acres—it is likely to support no more than 4,000 wells, and with its short production history, an optimistic projection of future Haynesville production would be 12 *tcf*.

3. **Fayetteville Shale:** The Fayetteville Shale is the second oldest major modern shale play in the US and the third largest by production.[62] With more nearly 4,000 wells drilled between 2005 and mid-2012 and approximately 3 tcf of production according to publicly available data from the Arkansas Oil and Gas Commission, we certainly have enough reliable production data to make a reasonable estimate of future recoveries.[63] Based on the core area that has shrunk to approximately 750,000 acres, a liberal estimate of future recoveries from the Fayetteville would be 7 *tcf*.

4. **Marcellus Shale:** Due to the limited production history of the Mar-cellus—the play has produced approximately 2 tcf as of mid-2012—re-

covery of an additional *50 tcf*, while not out of the question, is based on the assumption that much of what is now considered the core area of the play is commercially productive.[64]

5. **Woodford Shale:** The Woodford Shale of the Arkoma and Anadarko basins of Oklahoma produce on a combined basis approximately 600 mmcf/d. Since the State of Oklahoma is far and away the worst in the nation, among larger producing states, when it comes to providing production data to the public, my estimate of approximately 1 tcf of historical gas production from the Woodford Shale is based on the financial statements of the play's publicly traded operators. Commercial production was first established in the Arkoma Woodford in 2005 and has stabilized at approximately 300 mmcf/d. Commercial production from the Anadarko Woodford began in 2007–2008 and had grown to 300 mmcf/d by late 2011. A reasonable estimate for future recovery from the Arkoma and Anadarko Woodford shales would be *7 tcf*.

6. **Eagle Ford Shale:** The Eagle Ford Shale of south Texas is a rapidly growing unconventional resource play that came under commercial development in 2008. Since the Eagle Ford contains areas that are under development for oil production and liquids-rich natural gas, the play attracted great industry interest during the period of low natural gas prices between 2008 and 2012. Gas production in the play grew from practically nothing in 2008 to approximately 2 bcf/d by mid-2012.[65] While there is undoubtedly a significant amount of gas in the Eagle Ford Shale and production is likely to continue to grow due to high activity levels, much is still unknown due to its short production history. Therefore, due to the promising drilling results to date, a case can be made that the Eagle Ford may provide an additional *15 tcf* of production over its lifetime.

Minor and Emerging Shale Plays (Current Production < 500 mmcf/d)

1. **Antrim Shale:** Michigan's Antrim Shale is one of the oldest significant shale plays producing today. Largely developed during the 1980s and 1990s using vertical wells, it now offers more than two decades

of production data upon which to base an estimate of future production. According to a very detailed production history provided by the Michigan Public Service Commission, in the 21 years between 1989 and 2010 during which the play has been under commercial development, it produced almost exactly 3 tcf from approximately 10,000 wells.[66] The Antrim experienced peak average annual production of 546 mmcf/d, in 1998; by 2010 production had fallen 40 percent to an average of 330 mmcf/d.[67] It should be noted that production has fallen every year since 1998 despite the increase in gas prices between 2000 and 2008.[68] Given that the Antrim Shale has clearly entered a terminal and irreversible decline, a remaining resource of 2 *tcf* should be considered realistic.

2. Lewis Shale: Lewis Shale production has been established primarily in the San Juan Basin of New Mexico and Colorado. Most wells that produce from the Lewis also produce from either the Fruitland coal formation or the Mesaverde sandstone. Since most Lewis Shale production is commingled with that from other formations, reliable data is difficult to come by. In 1998, Burlington Resources—the largest operator in the San Juan Basin and later acquired by ConocoPhillips—attempted to produce from the Lewis Shale on a stand-alone basis but was unsuccessful in its efforts. According to a Society of Petroleum Engineers (SPE) paper published in 2000 (SPE 63091) titled "Lewis Shale, San Juan Basin: What We Know Now," a series of tests conducted by the authors determined that a well with a Lewis Shale completion can be expected to produce an additional 300 to 500 mmcf of gas.[69] Based on the higher of these two figures and the potential for 4,000 existing wells as well as future wells in the San Juan Basin to produce from the Lewis, an estimate of future recoveries of 2 *tcf* from the Lewis Shale is reasonable.

3. Avalon/Bone Spring Shale: The Avalon, also known as the Bone Spring or Leonard Shale, is a newly emerging play in Lea and Eddy counties of New Mexico. By mid-2011 natural gas production from the Avalon had reached approximately 100 mmcf/d and is expected to grow in the next few years.[70] Due to the limited publicly available drilling information to date, and the relatively small aerial extent of the play that has been identified so far, an aggressive estimate of future production for the Avalon/Bone Spring Shale would be 2 *tcf.*

4. Other Minor/Emerging Shales: Numerous other shale plays are currently under development in the US. Most have very little production history or the primary target is oil. For example, minor amounts of natural gas are currently produced from the Bakken, Mancos, Chattanooga, Mulky, Niobrara and Utica shales. While some of these plays have the potential to become significant gas producers, as do others in the exploratory stage, it is unclear whether this will ever happen. Therefore, assigning the minor/emerging shales a combined future production of *10 tcf* is generous at this stage of their development.

If we add up the likely future production of all shale plays under development, we get *132 tcf or about six years of supply* at current rates of consumption (approximately 24 tcf per year). This is an enormous amount of gas but a very far cry from the expected contribution of the shale promoters who have overstated the importance of shale gas.

But what about the shale plays that have yet to be discovered? Couldn't there be enormous resources and future production just laying in wait? Due to the pronounced importance of the first-mover advantage in the commercial development of shale plays—the first company to identify a new shale can acquire the best acreage—dozens of companies have spent billions of dollars over the past decade looking for the next Barnett. For example, since 2005, billions of dollars have been spent trying to crack the code of the New Albany, West Texas Barnett, Conasauga and Collingwood shales, all to no avail. No commercial production has been established in any of these plays and may never be. This should not come as a surprise to students of the oil and gas world. The biggest and most profitable fields are almost always found first. The biggest coal bed methane (CBM) field in the US is in the San Juan Basin. It was the first commercial CBM field in the US and remains not only the largest in the country but also the largest in the world. Likewise, the massive East Texas oil discovery was made early in the delineation of America's oil resources, and many of the largest natural gas fields in the Gulf of Mexico were found in the early days of offshore exploration.

In fact—and in direct contradiction to the claims about the bounty of shale gas in America and the rosy future it will provide us—it is entirely possible that all of our significant shale gas fields have already been discovered. Industry has spent billions of dollars looking for the next great American shale gas play and, thus far, there is little evidence to suggest there is another play the size of Barnett awaiting discovery.

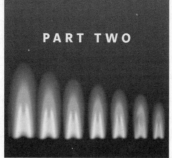

PART TWO

US NATURAL GAS DEMAND— AN INCREASINGLY INELASTIC DEMAND CURVE

SEVERAL CLEARLY identifiable trends in US natural gas demand have emerged over the past 15 years. As in the 1960s and 1970s, the major driver of demand growth has been the electricity-generating sector. The perception that natural gas will be cheap and plentiful for the foreseeable future, along with the view that it is environmentally advantageous to coal, has encouraged the electricity industry's increased consumption of natural gas. Essentially the industry has gone "all-in" on natural gas. However, unlike during 1970s natural gas crisis—which saw significant demand destruction in the electric power sector—mitigating any fall-off in deliverability will be far more difficult this time around. In addition to discussing trends in demand from the electric power sector in this section, I will also review the sticky demand from the residential and commercial building sector. Irrespective of economic activity or gas prices, America's enormous building stock needs to be heated in the winter and cooled in the summer.

Further, demand destruction from the industrial sector recently ended. After several decades of declining demand from America's industrial sector, nearly all of the forces that allowed for a significant portion of the US manufacturing base to move overseas are now in reverse.

Finally, I will examine the current consumption of natural gas by the country's fleet of natural gas vehicles (NGVs) and the outlook for NGVs to play a bigger role in our country's transportation system.

CHAPTER 6

Electric Power Industry Becomes Hooked on Natural Gas

THE ELECTRICITY-GENERATING sector has become the most important source of demand for natural gas in the US over the past 15 years. No industry has bet its future, and that of many Americans, more heavily on the myth of plentiful natural gas supplies. While electricity rates have remained subdued over the past couple of years, due to the recession after the financial crisis of 2008 and weak natural gas prices, the power industry's increased dependence on natural gas is very dangerous, both to our economy as a whole and to the immediate financial health of millions of Americans.

Simply put, rising natural gas prices will lead to higher electricity rates, since utility regulators set rates based on the highest cost form of power generation—which has historically been natural gas. (Nuclear generation is typically the lowest-cost source of power, followed by coal and natural gas.) To put the degree to which the industry depends on natural gas into perspective, consider the following: on the eve of the 1970s natural gas crisis, only 18 percent of US demand came from electricity generation, compared to over 30 percent today.[1] Therefore, the US was able to escape the grip of spiraling natural gas prices in the 1970s in part by reducing demand for natural gas from the electricity-generating sector. However, as reflected in Table 6.01 below, demand for

TABLE 6.01. US Electricity Industry Consumption of NG from 1970–1985 and 1994–2011.[2]

Year	Electricity industry NG consumption in tcf	Total NG US consumption in tcf	Electricity as percentage of total NG consumption
1970	3.932	21.139	18.60%
1971	3.976	21.793	18.24%
1972	3.977	22.101	17.99%
1973	3.660	22.049	16.59%
1974	3.443	21.223	16.22%
1975	3.158	19.538	16.16%
1976	3.081	19.946	15.44%
1977	3.191	19.521	16.34%
1978	3.188	19.627	16.24%
1979	3.491	20.241	17.24%
1980	3.682	19.877	18.55%
1981	3.640	19.404	18.75%
1982	3.266	18.001	17.92%
1983	2.911	16.835	17.29%
1984	3.111	17.951	17.33%
1985	3.044	17.281	17.61%
1994	3.903	21.147	18.45%
1995	4.237	22.207	19.07%
1996	3.807	22.609	16.83%
1997	4.065	22.737	17.87%
1998	4.588	22.246	20.62%
1999	4.820	22.405	21.51%
2000	5.260	23.333	22.54%
2001	5.342	22.239	24.02%
2002	5.672	23.007	24.65%
2003	5.135	22.277	23.05%
2004	5.464	22.389	24.40%
2005	5.869	22.011	26.66%
2006	6.222	21.685	28.69%
2007	6.841	23.097	29.61%
2008	6.661	23.226	28.67%
2009	6.887	22.834	30.16%
2010	7.387	23.775	31.07%
2011	7.601	24.309	31.27%

*Includes combined-heat-and-power plants (which is between 15% and 20% depending on year).

natural gas from this sector is now much larger, both quantitatively—up 3.66 tcf per annum, a 93 percent increase between 1970 and 2011—and as a percentage of total US demand.

Spurred on by spiraling prices and growing supply insecurity, the industry made a huge movement away from natural gas beginning in the 1970s. Between 1970 and 1978, annual consumption of natural gas for electricity generation *declined* by 744 bcf per annum.[3] This reduction was possible due to the rapid build-out of both nuclear and coal generating capacity. During this time, the number of nuclear reactors generating electricity in America grew from 20 to 70 and coal consumed in the electricity sector grew from 320.2 million short tons in 1970 to 481.2 short tons in 1978, a 50.3 percent increase.[4]

Today, though the percentage of electricity created from natural gas represents approximately the same portion of the pie as it did in 1970, the pie itself has gotten bigger—267 percent bigger—as America's appetite for electricity has grown.

High dependence on natural gas as a feedstock to generate electricity will be a big problem as US production declines gain momentum and imports become more difficult to secure. But couldn't we just have a replay of the 1970s, when coal and nuclear generating capacity took share away from natural gas? Unfortunately, no. Unlike then, now other sources of generating capacity will not be able to compensate for a decline in natural gas-fired generation. While there was an uptick in the start-up of coal-fired power plants in 2010—11 plants, the largest addition to the fleet since 1985, came online with 6,682 gigawatts of capacity—these recent additions are barely offsetting shutdowns of older, dirtier plants.[8] As a result of the perception that natural gas will remain cheap for years to come, more proposed coal-fired plants have been cancelled than announced in recent years, a strong indication that additions to the coal generation fleet are likely to be very limited.

Material growth in nuclear generating capacity is also difficult to foresee after the March 2011 meltdown of multiple reactors in Fukushima, Japan, and calls for increased scrutiny over the industry. Despite the closing of 22 nuclear reactors in the US since 1970, total nuclear electricity generation continues to trend higher in part because plant

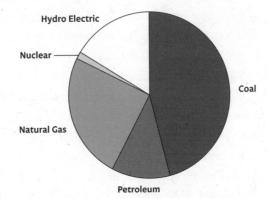

FIGURE 6.01. Electricity Generation Sources 1970.[5]

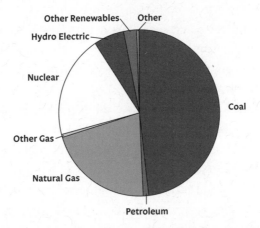

FIGURE 6.02. Electricity Generation Sources 2008.[6]

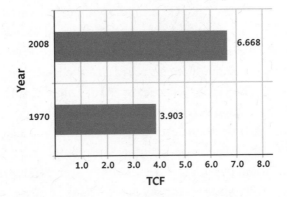

FIGURE 6.03. Natural Gas Consumption by Sector—Years 1970 and 2008 (tcf).[7]

turnarounds for maintenance have gotten shorter and shorter.[9] Uprating, which increases a reactor's output by refueling it with either slightly more enriched uranium or a higher percentage of new fuel, has also contributed to increased nuclear generation.[10]

However, safety concerns and increased plant downtime for maintenance will make it very difficult for the industry to achieve further efficiency increases. In fact, there is a strong possibility that increased downtime, rising maintenance costs and expensive repairs at a number of large plants will reduce output from the nation's nuclear plants for the next several years. According to a September 2012 Bloomberg article, Florida's 35-year-old Crystal River Unit 3 nuclear plant, eighty miles north of Tampa and out of service since 2009, will require repairs estimated at $1.3 billion to fix a cracked roof and other problems. The article also mentioned how the two 30-year-old San Onofre reactors near Los Angeles have been offline since January 2012 due to "leaks and unusual wear."[11] Additionally, in October 2012, Dominion announced it was closing its 556-megawatt Kewaunee nuclear plant in early 2013 due to weak power prices. The decision comes less than two years after the plant received a 20-year extension of its operating license from the Nuclear Regulatory Commission. Dominion will take a $281 million one-time charge related to plant closure costs.[12]

What about renewable energy? While I am one of its biggest fans and believe it will play a larger role in our energy future sooner than most think, the period of low gas prices and tighter capital markets that followed the financial crisis of 2008 slowed the rollout of many renewable projects. The falloff in electric generation from America's four thousand hydroelectric dams over the past fifteen years, due to the silting up of numerous dams and a lack of rainfall in certain areas, largely offset gains in wind and solar and other sources of renewable energy until very recently. Table 6.02 clearly displays this trend.

In conclusion, historically high reliance on natural gas by the electricity generating sector will make the upcoming deliverability crisis even more painful than that of the 1970s, since the economy will either have to shrink, become more efficient in its use of natural gas in generating electricity or build out other forms of generation.

TABLE 6.02. US Renewable Generation in Thousand Megawatt Hours.

Year	Hydroelectric generation	Other renewable generation*	Total renewable generation	Combined % of total generation
1996	347,162	75,796	422,958	12.28%
1997	356,453	77,183	433,606	12.41%
1998	323,336	77,088	400,424	11.06%
1999	319,536	79,423	398,959	10.79%
2000	275,573	80,906	356,479	9.37%
2001	216,961	70,769	287,730	7.70%
2002	264,329	79,109	343,438	8.90%
2003	275,806	79,487	355,293	8.94%
2004	268,417	83,067	351,484	8.85%
2005	270,321	87,329	357,650	8.81%
2006	289,246	96,525	385,771	9.49%
2007	247,510	105,238	352,748	8.48%
2008	254,831	126,212	381,043	9.24%
2009	273,445	144,279	417,724	10.57%
2010	260,203	167,173	427,375	10.36%
2011	325,074	194,993	520,067	12.67%

*Includes wind, solar photovoltaic, solar thermal, geothermal, wood, black liquor, other wood waste, biogenic municipal solid waste, landfill gas, sludge waste, agriculture byproducts and other biomass.[14]

CHAPTER 7

The Coming Increase
in Industrial Demand
for Natural Gas

W E ARE NOW at an important turning point for American demand for natural gas. For decades, Industrial America moved
operations overseas to take advantage of cheaper labor, low shipping
costs and cheap overseas gas. Part of the reason for the offshoring phenomenon can be explained by what *New York Times* columnist and author Thomas L. Friedman has famously described as the "flattening" of
the world economy. Flattening, more commonly referred to as globali-
zation—enabled by the Internet revolution as well as the breaking down
of trade barriers and advances in supply chain management—has had
a devastating impact on America's manufacturing base. As you can see
from Table 7.01, in the first decade of the new millennium, industrial
demand for natural gas in the US fell dramatically.

Between 2000 and 2009, industrial demand fell an incredible 24.25
percent to 6.167 tcf. How could the American economy grow during
a period when natural gas consumption dropped nearly 25 percent?
Simple. Practically all of our country's GDP growth during this period
was dependent on the now-burst finance/housing bubble and service
industry, rather than on manufacturing. In the ten years between 2000
and 2009, increasing numbers of American manufacturers shifted production overseas, taking their natural gas demand with them. *However,*

TABLE 7.01. US Industrial Natural Gas
Consumption 2000 to 2011.[1]

Year	Industrial consumption in tcf	Industrial consumption as % of total US consumption
2000	8,142	34.89%
2001	7,344	33.02%
2002	7,507	32.62%
2003	7,150	32.09%
2004	7,243	32.35%
2005	6,597	29.97%
2006	6,526	30.07%
2007	6,654	28.80%
2008	6,670	28.65%
2009	6,167	26.92%
2010	6,517	27.41%
2011	6,714	27.62%

this trend recently ended. In fact, the re-industrialization of America began in 2010 and will continue for decades, for five main reasons:

1. A Weaker US Dollar

At the turn of the millennium, the strong US dollar (USD) played a major role in shifting industrial production to foreign locations. In 2000, despite the popping of the technology and telecom bubbles, the USD enjoyed top billing amongst the world's major currencies. Japan was still trying to recover from its lost decade and the world had just been introduced to a new European currency. Ten years ago, the now-troubled European Monetary Union (EMU) was still finding its legs after having converted to a single currency in 1999. A weak euro gave German exporters a significant advantage over their American competitors. However Germany, whose export industries were also aided by the growth of the country's workforce after reunification in 1990, was not the only country whose exports benefitted from a strong USD. At the turn of the millennium, many Asian countries, still trying to recover from the Asian flu that devastated their economies and currencies in 1998, became a great place for American companies to locate manufacturing operations because most Asian currencies were still well below their pre-crisis levels.

However, by mid-2011, due to several years of record federal budget deficits and the S&P downgrade of the US federal government's "AAA" credit rating, the US dollar had fallen to post-WWII lows against the Japanese yen and was near a historic low compared to the euro. The fall in the dollar against the euro occurred despite the banking and fiscal troubles that continue to plague several members of the EMU. A weaker USD makes the US industrial sector more competitive in overseas markets and will likely boost natural gas demand from US manufacturers.

2. A Lack of Plentiful and Cheap Overseas Gas

Cheap natural gas in foreign locales encouraged large gas-consuming US manufacturers to shift production overseas. No sector has been more impacted by the lure of cheap foreign natural gas than America's nitrogen fertilizer industry. Natural gas is the primary feedstock in the manufacture of ammonia, which is used in turn to produce nitrogen fertilizers. According to the USDA, US nitrogen fertilizer capacity dropped 42 percent between 1999 and 2008 and annual production decreased 37 percent from 17.9 million tons to 11.2 million tons.[2] To compensate for this decline, the US turned to the import market in a major way. Imports surged 383 percent, from 2 million tons in 1999 to 9.7 million tons in 2008; and imports as a percentage of total supply grew from 12 percent to an *incredible 52 percent of all ammonium fertilizer spread on US farmland.*[3] Trinidad and Tobago has been a major beneficiary of the US's nitrogen production decline and now exports about a quarter of the nitrogen fertilizers used by American farmers. Canada, Russia and the Middle East have also increased their exports of nitrogen fertilizers to the US over the past decade. Transportation costs account for between 22 and 50 percent of the total cost of nitrogen fertilizers delivered to the Gulf Coast.[4] With few sources of cheap, "stranded" gas available anywhere in the world, and agricultural commodity prices continuing to trade at historically high levels, look for US fertilizer production to grow over the next five years.

One company that is certain to add to US natural gas demand for the manufacturing of fertilizer is Egypt's Orascom Construction Industries. In 2012, Orascom completed construction and commissioned a methanol and anhydrous ammonia facility in Beaumont, Texas, capable of producing 0.25 million metric tons of anhydrous ammonia and 0.75 million metric tons of methanol per year.[5] In September 2012, Orascom announced that its wholly owned subsidiary, Iowa Fertilizer Company (IFCo), plans to build a $1.4-billion nitrogen fertilizer plant in Wever, Iowa, close to the Mississippi River.[6] The plant is expected to begin operations in 2015. According a September 2012 Bloomberg article, "IFCo's new plant will be the first world scale natural gas-based fertilizer plant built in the United States in nearly 25 years and will help reduce the country's dependence on imported fertilizers which exceeds 15 million metric tons of ammonia, urea, and urea ammonium nitrate (UAN) annually."[7]

3. Increased Transportation Costs

Low transportation costs have helped put America's manufacturers under severe competitive pressure over the past twenty years. Massive cargo ships and cheap bunker fuel have allowed for the transportation of an increasing amount of bulk goods around the world at remarkably low costs. However, this trend is now reversing. Oil prices have more than doubled since the turn of the century and are poised to return to their 2008 levels, due to stagnating supply growth and increasing demand from fast-growing Asian economies. Therefore, the era of re-localization is now upon us. Former CIBC economist Jeffrey Rubin articulates this phenomenon very well in his book *Why your World is About to Get a Whole Lot Smaller*. In it, he provides several examples as to why higher shipping costs, due in large part to higher prices for bunker fuel (the fuel nearly all cargo ships require), will usher in an age in which American businesses are increasingly involved in manufacturing and consumers are once again buying local.[8] It only follows that increased local manufacturing will increase demand for natural gas from the industrial sector.

4. Increased Labor Costs

The allure of cheap labor, which had been a major factor in the decision of many manufacturers to relocate operations overseas, has lost much of its luster. In industries where automation has wiped out most gains from locating in areas of cheap labor, we will see manufacturing return to the US. More importantly, as Honda and other manufacturers have found out, workers in China and elsewhere are now demanding higher wages. A strike at a Honda plant in Foshan, China, in June 2010 shut down production for several days before management gave workers a $44 per month increase in wages and benefits, less than half of what they were seeking (currently, workers receive $117 per month plus housing allowances and bonuses).[9] Inspired by the success of the Honda strike, workers at two companies in the northern port city of Tianjin that supply parts to a Toyota production facility staged a strike that led to a brief shutdown of Toyota's production line.[10] Also in 2010, after more than a dozen workers committed suicide due largely to harsh working conditions, Foxconn, the world's biggest contract electronics manufacturer, announced a series of wage hikes that doubled the salaries of its more than 800,000 workers in China to 2,000 renminbi or nearly $300 a month.[11]

Given the rising food and high housing costs that are prevalent throughout Asia, I expect to see much more labor unrest over the next decade as workers increasingly demand better compensation. While Asian wages are still significantly lower than Japanese or Western rates, the gap will continue to narrow and make relocating manufacturing facilities overseas increasingly less attractive.

5. The Lack of Secure Intellectual Property

The final reason America's manufacturing base will not decline from current levels and is likely to grow substantially over the next decade is that manufacturers are growing wary of locating operations in countries that have little to no respect for intellectual property rights. For example, consider the experience that bankrupt hi-tech battery maker A123 Systems had in China. A123 creates some of the world's most technologically advanced lithium–ion batteries and its products can be found in electric and hybrid-electric vehicles as well as grid management systems. In a story that appeared in the *Chicago Tribune* on May 16, 2010, A123 management discussed their reluctance to locate additional manufacturing facilities in China due to concerns over the theft of their patents.[12]

A123 Systems cofounder and Chief Technology Officer Bart Riley cited lack of respect for intellectual property rights as a driving force (along with a $250 million grant from the US DOE) in the company's decision to build a new battery plant in Michigan and develop plans for additional factories in the US rather than in China. Manufacturing cutting-edge technology in China has its negatives, as Riley noted in the *Chicago Tribune* story:

> "But in ramping up production in China, A123 paid an immeasurable price," Riley said.
> "The company did what it could to slow the technology transfer by breaking down the manufacturing process into steps," Riley said, but "we ended up having to teach these guys how to make our state-of-the-art, world-class batteries.... And some of them are (now) competing with us directly."[13]

Rebounding industrial natural gas demand in 2010, 2011 and the first half of 2012 is a clear sign that America's industrial sector is gaining strength and is likely a harbinger of things to come. A weakening US dollar, lower wages due to structurally high US unemployment and increasing automation technology, combined with increased shipping costs, are once again making the US manufacturing sector competitive.

CHAPTER 8

Commercial and Residential Consumers Add to Demand Inelasticity

D ESPITE GROWTH in America's residential and commercial real estate sector over the past three decades, demand for natural gas to heat and cool America's buildings has grown only modestly. In 2011, 7.89 tcf of natural gas went towards heating and cooling America's building stock—approximately 32 percent of total US consumption.[1] Further, as America's building stock grew, demand for electricity increased. As I discussed in Chapter 6, the electricity-generation industry has become increasingly reliant on natural gas-fired power plants for its production. Figure 8.01 displays how electricity demand from America's buildings grew significantly in the years between 1980 and 2005, while demand for natural gas remained fairly constant.

According to a 2008 study by the US Department of Energy (DOE), the number of US households went up 40 percent—from 80 million to 113 million—between 1980 and 2005.[3] This increase was largely driven by population growth, from 228 million residents to 300 million; however, as Table 8.01 shows, residential demand during this period barely grew.[4] Meanwhile, demand to heat and cool commercial buildings rose 14.5 percent, largely because of a doubling of US GDP in real terms (2005 dollars), from $5.8 trillion to $12.4 trillion.[5] As you can see in the table, commercial demand has flattened since 2000.

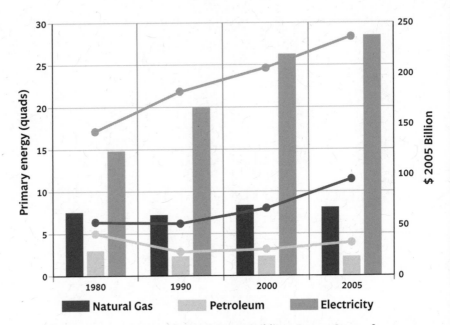

FIGURE 8.01. Predominance of Electricity as Buildings Energy Source.[2]

TABLE 8.01. US Heating and Cooling Demand.[6]

Year	Residential demand in tcf	Commercial demand in tcf	Total in tcf
1980	4.75	2.61	7.36
1985	4.43	2.43	6.86
1990	4.39	2.62	7.01
1995	4.85	3.03	7.88
2000	4.99	3.18	8.17
2005	4.82	2.99	7.81
2010	4.78	3.10	7.88

Source: EIA.

Looking forward, I expect demand for natural gas to heat and cool America's residential and commercial buildings to remain stable for the next decade or so, for a couple of reasons. First, the collapse of the real estate bubble in 2005–2006, and the huge inventory of unoccupied homes and commercial buildings that continues to plague America's cities and towns, will almost certainly prevent our country's building stock from expanding for the foreseeable future. However, much to the chagrin of America's banks—who are now among the country's biggest landlords—most homes and commercial buildings, whether occupied or not, either need to be heated in the winter to prevent pipes from freezing or air-conditioned nearly year-round to prevent mildew from destroying their value. Second, high vacancy levels, falling property values and tightness in the credit markets will keep many commercial landlords and homeowners from making the investments required to increase energy efficiency. Only a combination of substantially higher natural gas prices, increased tax incentives and the obsolescence of older, inefficient buildings will result in a material reduction of demand for natural gas from the sector.

There should be little doubt that renewable methods for heating and cooling buildings, such as solar panels and geothermal systems, are likely to grow substantially from today's levels as technology improves and natural gas prices rise. However, despite generous federal tax breaks and other incentives, these systems are installed in a very small percentage of America's building stock.

CHAPTER 9

Natural Gas Vehicles: A Non-Starter

THOUGH MUCH has been made about the potential for natural gas vehicles (NGV) to reduce both motor vehicle emissions and our dependence on foreign oil, they are unlikely to be adopted on a wide scale in the US for a number of reasons. I see the push towards increased adoption as a profoundly flawed policy, since any large increase of NGVs is predicated on the false belief that natural gas will remain cheap and abundant for decades.

NGVs first appeared on America's roads shortly after the passage of the Energy Policy Act of 1992, which required fleets operated by federal and state governments to replace between 10 and 30 percent of their conventional motor vehicles with alternative fuel vehicles (AFVs) between 2000 and 2010.[1] It should be noted that because the Policy Act of 1992 did not specifically require the use of natural gas in AFVs, fleet operators also purchased a significant numbers of flex-fuel vehicles—which use a blend of gasoline and ethanol—to comply with its requirements. NGV fleet vehicles (mostly buses and a few garbage trucks) began appearing on city streets in 1994 and currently make up the vast majority of NGVs on America's roads. The federal government, initially a major buyer of NGVs, has increasingly purchased flex-fuel vehicles in recent years to meet its AFV requirements and currently operates only slightly more than 12,000 NGVs out of a total fleet of approximately 85,000 vehicles.[2]

Despite substantial policy efforts to encourage their adoption, NGVs represent a tiny fraction of all vehicles on US roads today. According to the International Association for Natural Gas Vehicles (NGVA), as of December 2010, there were approximately 112,000 NGVs in service.[3] Even factoring in substantial growth in their adoption in the two years since the study was completed, vehicles powered by natural gas still account for far less than 1 percent of the estimated 250 million registered vehicles on US roads today.[4] Surprisingly, even though the amount of natural gas consumed by NGVs has steadily increased over the past 15 years (see Table 9.01)—from 8.3 bcf per annum in 1997 to 32.8 bcf per annum in 2011—the number of refueling stations has *dropped*, from 1,400 to slightly fewer than 1,000 during this period.[5] Nearly all refueling stations in use today in the US are for the support of fleet vehicles, with only a small fraction open to public use. To compare, there are approximately 164,000 refueling stations in the US for gasoline and diesel alongside America's roads and highways.[6] Another headwind facing wider adoption of NGVs was the passage of the 2008 National Defense Act, which designated hybrid electric vehicles (HEVs), such as the Toyota Prius, as eligible to meet AFV requirements.[7] The more choices

TABLE 9.01. NGV Demand 1997 to 2011.[8]

Year	US natural gas vehicle fuel consumption (mmcf)
1997	8,328
1998	9,341
1999	11,622
2000	12,752
2001	14,536
2002	14,950
2003	18,271
2004	20,514
2005	22,884
2006	23,739
2007	24,655
2008	25,982
2009	27,262
2010	30,670
2011	32,850

Source: EIA

federal agencies have to meet their AFV goals, the less likely it is that the government will reverse course and embrace NGVs.

Another reason NGVs have not caught on is the high cost of building fueling stations. With an estimated initial cost of $350,000 per station, promoters of NGVs have found few in the private sector willing to make the financial commitment to building out the needed infrastructure without substantial tax credits.[9] Perhaps unsurprisingly, the Natural Gas Act (H.R. 1835 and S. 1408), is loaded with them. While the House bill (H.R. 1835) differs slightly from the Senate's version (S. 1408), here are a few of the highlights from the two bills:

- Tax credit of 50 cents per gasoline-gallon-equivalent of CNG or liquid gallon of LNG vehicle fuel (equivalent to $0.57 per diesel gallon for CNG and $0.87 per diesel gallon for LNG)
- Tax credit of up to 80 percent of the vehicle's incremental cost for all dedicated natural gas-fueled vehicles (up to a maximum credit of $64,000 per vehicle)
- Tax credit of up to 50 percent of the vehicle's incremental cost for bi-fuel natural gas-fueled vehicles (up to a maximum credit of $64,000 per vehicle)
- Tax credit of up to $100,000 per station (CNG, LNG or LCNG)[10]

Though it is unclear when, if ever, the Natural Gas Act will make it out of Congress, its primary backer in the private sector, T. Boone Pickens, has vowed to keep the pressure on.[11]

I believe a far better way to reduce our consumption of oil for transportation purposes is to embrace electric vehicles (EVs). (Full disclosure: as of this writing in the fall of 2012 I do not have any EV-related investments or financial connection to the industry.) EVs on the road today span a wide range of sizes and uses. From cars (such as the Tesla Roadster and Model S, the Nissan Leaf and the Mitsubishi iCar) to municipal buses and trucks already in service, not to mention dozens of new, all-electric models of every stripe likely to be rolled out between 2013 and 2015, America seems ready to embrace the modern electric vehicle. Given the tremendous advances in lithium-ion battery technology over the past several years, it is increasingly clear that electric vehicles are becoming the alternative to the internal combustion engine-powered vehicle.

A very important part of the unfolding EV success story is the huge number of recharging stations already in place and the low cost of new stations. According to the US Department of Energy, as of December 2012, there were more than 5,000 public EV recharging stations already installed across the nation as well as many more private ones.[12] Level II charging stations, often found in public garages or parking and capable of charging an EV in three to six hours, can be installed for between $4,000 and $9,000, while DC charging stations, which will charge a car in as little as thirty minutes, can be installed for approximately $25,000 each.[13] Additionally, nearly every electrical wall socket in America is capable of recharging an EV with only a modest investment (usually $1,000 or so) in a home recharging station. It should be noted that even the most expensive public EV charging station is approximately *one-tenth* the cost of a NGV refilling station, while the cheapest public EV stations can be installed for only *two percent* of the cost of a typical NGV refilling station. Several companies, including General Electric and Siemens AG appear poised to make EV charging stations available to most Americans in a few years' time.

The biggest reason I favor EVs over NGVs is that EVs can be powered from renewable sources of electricity. I am well aware that there is not enough solar or wind generation capacity installed today to power even a fraction of the cars on the road. However, should electricity and gasoline prices rise sufficiently, a combination of solar panels and energy storage devices—to allow for charging when the sun is not shining—is likely to become a common sight in neighborhoods throughout America. Google, which has a history of spotting trends, already has a solar-powered electric car re-charging carport in its parking lot, and I expect many other companies to set up similar venues in the near future. One company that is at the forefront of the technology is GE, which has announced a partnership with Inovateus Solar for the building of EV solar carports that will be capable of charging between two and 13 EVs per day.[14]

Lastly, further adoption of NGVs at a time of tightening natural gas supplies is simply trading one hydrocarbon crisis for a potentially more damaging one. While I fully agree with NGV supporters who argue that importing nearly 10 million barrels of crude and refined products on a daily basis is damaging the US economy and unsustainable in the long term, increased reliance on natural gas for transportation is a dead end. The US imports oil from numerous countries, and, should imports from one country decline, new relationships can be developed to replace any shortfall. In the case of natural gas, the US depends almost solely on imports from Canada, a country that is currently obligated under the NAFTA to continue exporting gas in proportion to its production.[15] However, former Princeton University professor, M. King Hubbert disciple, and author, Kenneth S. Deffeyes, suggested in his book *Beyond Oil*, that Canada's obligation to export natural gas under NAFTA is unlikely to continue should the country develop a natural gas crunch:

> Although Canada, the United States and Mexico are all major gas producers, Canada exports gas to the United States and the United States exports gas to Mexico. NAFTA requires all three countries to continue their historic pattern of sharing gas. During a serious gas crunch, however, *NAFTA might get wadded up and pitched in the wastebasket.* [Emphasis added.][16]

Given Canada's growing consumption of natural gas due to its strong economy, growing oil sands production (a big consumer of natural gas), ability to export LNG to Asia beginning in 2014 and dramatic fall-off in production since 2002, there is a high probability that exports to the US will drop to zero within the next decade. (I will discuss Canada's supply outlook further in Part Three.) Additionally, despite the US's significant LNG import capacity, we will find it very difficult to import meaningful amounts of LNG once the deliverability crisis begins due to a lack of long-term supply contracts. Without the ability to import increasing amounts of natural gas once production in the US drops off significantly, we would potentially see a disruption in our transportation network. In addition to the hardships that Americans faced in the natural gas crisis of the 1970s, an increased reliance on NGVs during the coming deliverability crisis will make a bad situation worse.

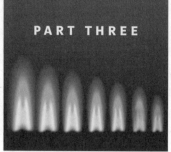

PART THREE

SUPPLY: WE WILL NOT PRODUCE OUR WAY OUT OF THE NEXT CRISIS

JUST AS NATURAL GAS production peaked in 1973 and in 2001, the US reached another inflection point in 2011.[1] In that year, the US Energy Information Administration estimated that the country reached its third natural gas production peak, at 66.2 bcf/d, a figure that will likely never be materially exceeded due to falling conventional production and a flattening of shale gas production growth.[2] Part Three of this book is devoted to reviewing America's most important sources of natural gas. I take an in-depth look at five of the country's largest gas-producing states and the Gulf of Mexico, which account for more than 73 percent of the country's production.[3] I will also explore the dichotomy of America's producing basins. Over the past four years nearly all production growth has come from a handful of shale gas fields while nearly every other field has entered terminal decline.

No review of supply would be complete without examining the importance of imports to America's overall supply picture. I will review how America's years of significant reliance on foreign sources of supply is about to end at the worst possible time, right as the country's productive capacity is set to fall materially. More specifically, Canada will soon end its multi-decade-long practice of exporting cheap natural gas to the US due to falling production, increased domestic demand and new liquefied natural gas (LNG) export facilities under development in British Columbia designed to tap the lucrative Asian LNG market.

Lastly, I will review the state of the global LNG market. More than a decade ago, the US built substantial LNG import capacity in the misplaced belief that it could secure cargoes on the spot market. Despite the increase in import capacity, the US was unable to secure significant LNG imports to alleviate the 2008 price spike. More importantly, during the upcoming supply crunch, it is highly unlikely that the US will be able to contract significant additional supplies of LNG due to tightness in the world market.

CHAPTER 10

Texas:
The Big Enchilada

I T IS DIFFICULT to overstate the importance of Texas to the US natural gas supply chain. The state is by far the largest and most important producer of natural gas in the country. In fact, Texas produces 29 percent of all natural gas in the US on a daily basis.[1] If Texas were a country—and many of its residents believe it is one—*it would be the third largest producer of natural gas in the world behind Russia and the rest of the US.*

The Texas Railroad Commission, the state agency responsible for petroleum data collection, divides the State into 10 districts and breaks down production data along district lines. Figure 10.01 is a graphical display of the state's districts.

District 1

District 1 is home to a significant portion of the Eagle Ford Shale, which has seen explosive growth in oil and natural gas production over the past five years. The Eagle Ford was first discovered in the late 1970s while industry players were developing the Austin Chalk—which sits directly above the Eagle Ford—and Edwards formations. Due to the high carbonate content of the shale—up to 70 percent—it is very brittle and very amenable to hydraulic fracturing.[4] The Eagle Ford Shale, which lies between four and twelve thousand feet below the surface, is roughly

TABLE 10.01. World's Largest NG Producers in 2011.[2]

Country	Production in tcf
US (includes TX)	22.99
Russia	21.43
Texas	7.60
Canada	5.67
Iran	5.36
Qatar	5.18
China	3.62
Norway	3.58
Saudi Arabia	3.50

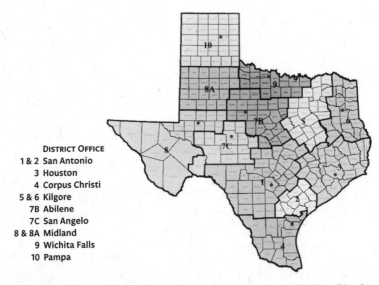

DISTRICT OFFICE
1 & 2 San Antonio
3 Houston
4 Corpus Christi
5 & 6 Kilgore
7B Abilene
7C San Angelo
8 & 8A Midland
9 Wichita Falls
10 Pampa

FIGURE 10.01. Railroad Commission of Texas—Oil and Gas Division District Boundaries.[3]

fifty miles wide and four hundred miles long and is the source rock for the Austin Chalk and the giant East Texas Oil Field.[5] Figure 10.02 shows how there are three distinct "windows" within the Eagle Ford; the oil window, the wet gas/condensate window and the dry gas window.

Developers of the play benefit both from the area's topography and the existing gas-gathering and -processing infrastructure that was built during the development of the Austin Chalk play. Several new natural gas pipelines planned for the play will boost take-away capacity over the next several years.[7]

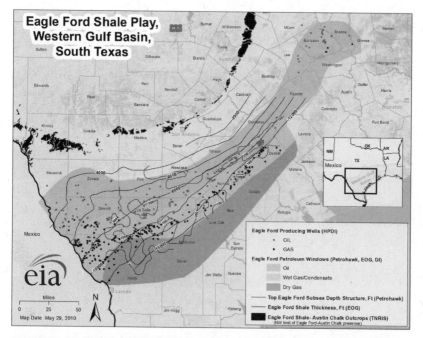

FIGURE 10.02. Eagle Ford Shale Play—Location, Wells and Petroleum Windows.[6]

While the Eagle Ford is undoubtedly one of America's brightest oil prospects and will be the center of industry activity for years to come, the play's natural gas deliverability remains uncertain. Though many very productive wells—in excess of 3 mmcf/d—have been drilled in both the dry gas window and the wet gas window, the play has a limited history of gas production. According to the Texas RRC, from 2004 through September 2012, the Eagle Ford had produced 900 bcf of natural gas from both gas wells and oil wells (casinghead gas).[8] It should be noted that while the Eagle Ford has produced since 2004, the vast majority of drilling in the play has occurred since 2008. While the Eagle Ford is a very meaningful unconventional oil and gas discovery that is likely to grow for years to come, it has yet to make District 1 a large producer of natural gas. In 2011, the district accounted for approximately *four percent of gas production in the State of Texas.*[9] District 1 production grew from 0.25 bcf/d in 2008 to 0.84 bcf/d in 2011, a 234 percent increase.[10]

Districts 2, 3 and 4

Districts 2, 3 and 4, located along the Gulf Coast, have experienced steep production declines since the peak of the natural gas drilling frenzy in late 2008. Production from the three districts fell from a combined 6.03 bcf/d in 2008 to only 4.25 bcf/d in 2011, a 30 percent drop.[11] The districts' major producing formations—the Vicksburg, Frio and Wilcox zones—are Tertiary-aged sandstones found between nine and sixteen thousand feet. All three have high porosities (9–20 percent) and low permeabilities (in the 0.01 to 1 md range) and wells drilled into them typically have high initial production rates followed by very steep declines.[12] The Vicksburg, Frio and Wilcox zones have been developed through vertical drilling for more than fifty years.[13]

I find the fate of districts 2, 3 and 4 very representative of the future of many other conventional, vertically developed areas throughout the country. In 2012, vertically developed conventional onshore fields provided the US with approximately 45 percent of its gas production. With nearly all of the publicly traded natural gas-focused E&P companies (and many privately held companies as well) favoring the horizontal development of unconventional gas resources (mostly shale but some tight sands), areas such as districts 2, 3 and 4 have been neglected. While their remaining deliverability potential is unclear, there should be little doubt that their best days are behind them.

District 5

Even before the discovery of the Barnett Shale, District 5 had a long history of oil and gas production. The first major field in the district, the Boonsville Field in Wise and Jack counties, was discovered in 1945 and has produced an estimated 3.1 tcf to date, from the Bend Conglomerate.[14] In addition to this significant field, operators in District 5 have established dozens of smaller fields targeting conventional formations such as the Morrow and Atoka sandstones and the Marble Falls limestone. It should be noted that the Texas RRC considers all Barnett Shale production to be part of the Newark East Field that counts towards District 9 production totals. Therefore all Barnett Shale production in Jack and Wise counties goes towards District 9 totals.

Due largely to the maturity of gas fields in District 5, production fell from 2.14 bcf/d in 2008 to 1.75 bcf/d in 2011.[15] Further declines are expected unless new discoveries are made.

District 6

Located in eastern Texas, District 6 is home to the Texas portion of the Haynesville Shale play as well as numerous large unconventional fields that have produced trillions of cubic feet of gas—mostly from Cotton Valley sands—over the past several decades. Despite the development of the Haynesville, production in District 6 between 2008 and 2011 grew from 3 bcf/d in 2008 to 3.2 bcf/d in 2011, a rise of only 7 percent.[16]

District 8

District 8 has seen a modest decline in gas production over the past several years, despite a drilling boom in the Permian Basin of west Texas. According to service company Baker Hughes (publisher of the Baker Hughes rig count), the number of active rigs in the district has risen from 120 in January 2008 to 285 in December 2011, an increase of 137 percent.[17] Operators have flocked to the area to apply horizontal drilling and modern fracturing techniques to the large potential of the oil-prone Wolfcamp Shale and Spraberry Trend among other targets. Between January 2008 and December 2011, oil production in District 8 grew from 9.8 million barrels per month to approximately 12.9.[18]

District 8 is unique among the reporting districts in Texas in that it is the only one where production of casinghead gas—gas produced from an oil well—exceeds gas well gas production. Casinghead gas production in District 8 increased from 0.9 bcf/d in 2008 to 1.16 bcf/d in 2011, a 29 percent rise.[19] However, due to a decline in production from gas wells, which dropped from 1.18 bcf/d in 2008 to 0.8 bcf/d in 2011, overall production fell from 2.08 bcf/d to 1.96 bcf/d during this period.[20] Due to few new gas finds in recent years, I would expect District 8 gas production to stay relatively flat for the next several years as gas from oil wells offsets declining production from gas wells.

Lastly, declining gas production in District 8 between 2008 and 2011 dismisses one widely held myth about the importance of casinghead gas to America's supply. Some observers believe that gas produced along with well oil production will significantly increase America's gas deliverability. Despite a huge uptick in drilling and oil production in Permian Basin since 2008, average daily gas casinghead gas production increased only 0.26 bcf/d by 2011. Similarly, only a modest amount of natural gas is produced along with oil from the Bakken formation of North Dakota and Montana despite the huge increase in oil production. Therefore it is safe to posit that gas produced along with oil production, whether it be in the Permian Basin or the Williston Basin of North Dakota and Montana or elsewhere, will have only a token impact on America's gas supply.

District 9

Discovered in 1981, the Barnett Shale of District 9 has been a remarkable success story that has dramatically changed the oil and gas industry.[21] Officially called the Newark East Field by the Texas Railroad Commission, the Barnett has produced more shale gas than any other field in the world and was the laboratory where George P. Mitchell, the father of the modern shale gas industry, spent more than a decade improving existing technology (hydraulic fracturing) to extract gas from shale. Figure 10.3 shows how production exploded in District 9's Newark East Field shortly after the turn of the millennium once hydraulic fracturing was implemented on a wide scale.

The Barnett Shale, which has had more than sixteen thousand wells drilled into it to date, has been responsible for the huge growth in District 9 in recent years.[23] Total production in the district has grown from 2.15 bcf/d in 2006 to 5.78 bcf/d in 2011, a 168 percent increase.[24] However, production growth from the field has slowed materially in recent years due to its increasing maturity, low gas prices and the increasing focus on the northwestern part of the field, where wells can be expected to produce more higher-value condensate and oil but a smaller amount of natural gas than wells in the dry gas portion of the play.

With the Barnett accounting for approximately 31 percent of natural gas production for the state, the play's future production will have a very large impact on production from Texas.[26] Given the very large number of wells already drilled into the Barnett, the steep decline curve of new

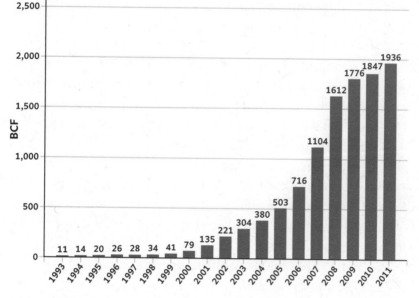

FIGURE 10.3. Newark East (Barnett Shale) Gas Well Gas Production (1993–2011).[22]

wells and the increased focus of new drilling efforts on the more liquids-rich part of the play, I suspect 2011 will mark the high-water mark for natural gas production from both the Barnett Shale and District 9.

District 10

Located in the Texas Panhandle, District 10 has seen activity increase significantly in recent years due to the application of horizontal drilling technology to the Granite Wash formation. The Granite Wash is a tight sands horizon that was first developed vertically in the 1950s and is currently being developed horizontally with great success. Operators such as Cimarex Energy (NYSE:XEC), Linn Energy (NASDAQ:LINE), Newfield Exploration (NYSE:NFX), Chesapeake Energy and Forest Oil (NYSE:FST) have used horizontal wells and multi-stage fracturing to bring new life to the Panhandle. Similar to the Eagle Ford Shale, activity in the Granite Wash has remained strong despite the weak gas prices of recent years due to the high liquids content of the natural gas production stream. For example, in July 2010, Linn Energy announced that its second horizontal well from its Stiles Ranch project area in the

Panhandle came onstream for the first 24 hours at a rate of 27 mmcf/d and 3,190 barrels a day of condensate.[27]

More recently, in June 2012, Chesapeake Energy announced excellent production rates from the company's first two wells from the Hogshooter formation. The Hogshooter is shallower and more oil prone than the Granite Wash and is found in largely the same area of western Oklahoma and the Texas Panhandle. One well averaged 6,600 boe/d of liquids production along with 4.6 mmcf/d of gas over its first eight days while another well averaged 1,665 boe/d of liquids and 1.4 mmcf/d of natural gas over its first 27 days.[28] While both of these wells produce a meaningful amount of natural gas along with a tremendous amount of liquids, the Hogshooter play is known to be fairly small. Only a modest number of future drilling locations will be developed over the next few years, so it is unlikely the play will ever produce substantially more than 100 mmcf/d over its lifetime.

Despite the uptick in drilling activity in recent years and the tremendous success of horizontal drilling into the liquids-rich Granite Wash in north Texas, natural gas production in District 10 grew from 1.69 bcf/d in 2008 to only 1.74 bcf/d in 2011, a 3-percent increase over the course of four years.[29]

Statewide Production, 1970 through 2011

Despite all of the advances in technology that have brought into commercial production the Barnett, Haynesville and Eagle Ford shale plays, the Granite Wash and Cotton Valley sands, and hundreds of other fields over the past four decades, the Lone Star state has never matched its peak production level: 26.3 bcf/d way back in 1972.[30] The closest it has come was in 2008 when gas production reached 21.44 bcf/d, fully 21 percent below its prior high.[31] More importantly, the 2008 peak was reached with 96,502 active natural gas wells while the 1972 summit was achieved with only 23,373 wells.[32] *In other words, it required more than 300 percent increase in the number of active wells to produce 21 percent less gas.* Figure 10.04 shows the rising number of wells in the state—which reached an all-time high of 101,831 in 2011—and the fluctuations in natural gas production over more than four decades.

FIGURE 10.04. Texas Producing Wells and Production 1960–2011.[33]

Lastly, Texas is a classic lesson in how geology trumps technology. Similar to the countless old, large oil fields, such as Mexico's Cantarell and California's Kern River, that saw a temporary bump in production after the application of secondary recovery technology, the incredible technology that has been deployed in the gas fields of Texas can no longer hold back natural depletion. With natural gas drilling activity in the state down substantially from 2008 peak levels, look for production declines for America's largest producing state to accelerate until at least 2015.

CHAPTER 11

Louisiana's Fleeting Production Rebound

AFTER REACHING peak natural gas production in 1970 of slightly more than 15 bcf/d, production in the State of Louisiana—the second biggest natural gas producer in the Lower 48—went into a 35-year downward spiral that reached its lowest point in 2005.[1] In that year, the state saw production bottom out at only 3.5 bcf/d, an incredible drop of 77 percent from the 1970 peak.[2] However, advances in drilling and completion technologies opened up several new horizons in the Bayou State and breathed new life into its gas industry. More specifically, the development of the tight, gas-bearing sands of northwest Louisiana and the Haynesville Shale, also located in the northwestern corner of the state, not only arrested Louisiana's multi-decade production decline but led to a 138 percent increase in output in only six years.[3] However, reduced activity in the Haynesville and the high rate of decline of many of the state's wells meant that by the end of 2011, production in Louisiana had flattened out and was rolling over by mid-2012.

Louisiana has a very long history as an important natural gas-producing state. Natural gas was first discovered in the state, by accident, in 1870, by a night watchman for the Shreveport Ice Factory after lighting a match near a recently drilled water well. The ice factory later used its newfound energy source as illumination for its plant.[4] In 1908,

Shreveport became one of the first Southern cities to have gas service after a pipeline was laid to connect the town to the prolific Caddo Field.[5] Though fields in the northwestern portion of the state provided the majority of early production, by the early 1950s, southern Louisiana was experiencing a production boon due to the discovery of numerous prolific fields. It was these new discoveries that supplied the feedstock for the rapid growth of America's chemical and fertilizer industries in the 1950s and 1960s. However, no boom lasts forever. Production from southern Louisiana peaked at over 12 bcf/d in the early 1970s and then entered a terminal decline as existing fields matured and were not supplemented by new discoveries.[6] In 2011, onshore southern Louisiana produced slightly more than 1 bcf/d, a 92 percent drop from its early 1970s peak.[7]

One of the main drivers behind the falloff in natural gas production in southern Louisiana is the high-pressure, fast-depleting nature of its producing formations. Southern Louisiana geology is dominated by the Gulf Coast Salt Dome Basin, which is characterized by deep (typically

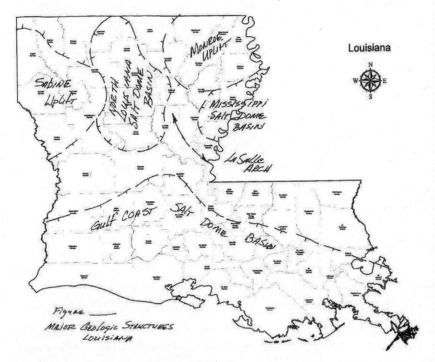

FIGURE 11.01. Map of the Salt Dome Basin.[9]

Salt is a peculiar substance. If you put enough heat and pressure on it, the salt will slowly flow, much like a glacier that slowly but continually moves downhill. Unlike glaciers, salt which is buried miles below the surface of the Earth can move upward until it breaks through to the Earth's surface, where it is then dissolved by ground- and rain-water. To get all the way to the Earth's surface, salt has to push aside and break through many layers of rock in its path. This is what ultimately will create the oil trap.

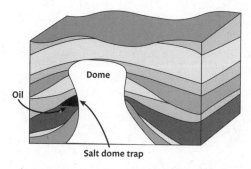

Here we see salt that has moved up through the Earth, punching through and bending rock along the way. Oil can come to rest right up against the salt, which makes salt an effective trap rock. However, many times, the salt chemically changes the rock next to it in such a way that oil will no longer seep into them. In a sense, it destroys the porosity of a reservoir rock.

Salt dome trap

FIGURE 11.02. Salt Dome Trap.[11]

12,000 feet), high-permeability, hydrocarbon-bearing structures created through the upward migration of salt during the creation of Louisiana's land mass.[8] Figure 11.01 shows the outline of the Salt Dome Basin.

Early operators exploring for oil found it easy to identify and exploit hydrocarbon structures adjacent to salt domes with the prospecting technology of the era, such as the torsion balance and the seismograph.[10] Trillions of cubic feet of natural gas were also found in salt dome structures. Figure 11.02 shows how the mobilization of salt created many of the hydrocarbon traps of southern Louisiana.

Offshore Louisiana

Natural gas production in Louisiana's offshore area is likely to continue its terminal decline due to its advanced maturity. Production off the state's waters has fallen from a high of 1.7 bcf/d in 1973 to only 191 mmcf/d in 2011, a drop of 89 percent. However, the application of advanced seismic, drilling and completion technology has the potential to open up the very deep horizons that sit below a thick layer of salt (often referred to as a salt weld and commonly found at approximately 25,000 feet) that could hold trillions of cubic feet of gas reserves. Due to the extremely high pressures and temperatures found in formations

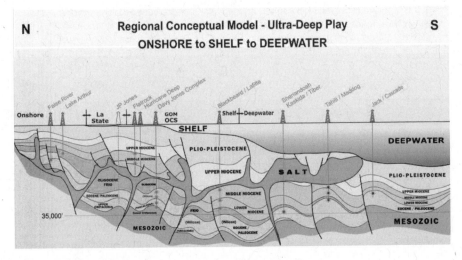

FIGURE 11.03. Regional Conceptual Model—Ultra–Deep Play.[12]

below the salt weld, operators are having to develop new methods for testing and completing these ultra-deep wells. The company that is leading the charge into deep horizons in Louisiana's offshore region is McMoRan Exploration (NYSE:MMR). Figure 11.03 is a graphical display of the various water depths in which the company operates and the numerous targets it is pursuing all the way down to 35,000 feet.

Though MMR has been in the news over the past few years with its ultra-deep (and very expensive) Blackbeard and Davy Jones wells in federal waters, the company is also pursuing deep natural gas prospects in the shallow state waters off the coast of Louisiana (federal waters begin approximately three miles from shore)[13] as well as onshore. The company has succeeded in developing a few above-the-weld deep prospects onshore but has not had much success in state waters so far.[14] While the deep horizons of Louisiana's shallow waters, both above the salt weld and subsalt, may hold vast potential, MMR and other operators are still in the early stages of exploration and it will be years before the full potential of deep horizons can be quantified. Lastly, though there has been much excitement in the media over the potential of sub-salt resources, as of late 2012, no company has been able to commence production from a productive formation below the salt weld either onshore or in the Gulf of Mexico.

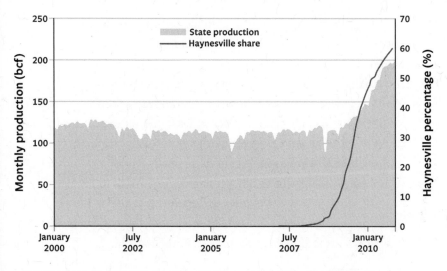

FIGURE 11.04. Haynesville Contribution to Louisiana Gas Production.[15]

North Louisiana

Since the commercialization of the Haynesville Shale in 2007–2008, the play has been the focal point of industry activity on the Sabine Uplift, an area of prolific oil and gas production for more than a century. By mid-2009, this play had become the driving force in the state's production growth.

Due to the high deliverability of many early Haynesville wells—some have initial production rates of over 15 mmcf/d—and the aggressive promotion of the play by several of early entrants, expectations for the Haynesville grew quite high. Though Haynesville wells are more costly than other shale wells, approximately $10 million each due to the depth of the formation, they are also expected to have higher recoveries per well. According to a December 2010 *Oil and Gas Journal* article, PetroHawk Energy, one of the most active early players in the play, estimated that each Haynesville well could be expected to recover 10 bcf, or nearly four times what the industry would expect to recover from a Barnett Shale well.[16]

One early notable promoter of the Haynesville was Chesapeake Energy CEO Aubrey McClendon. Here is an excerpt from a Platts' article that appeared on the Rigzone website in February 2009:

The Haynesville Shale may eventually become the world's largest producing gas field, Aubrey McClendon, CEO of Chesapeake Energy and a pioneer of the play in east Texas and northwest Louisiana, said Wednesday. Chesapeake expects the play, which only became widely known when the company began talking about it last March, will produce at least 500 tcf over time and then recover around 700 tcf before potentially growing even larger, McClendon said during a presentation to the annual Cambridge Energy Research Associates conference in Houston.

"We think in time it will become the largest gas field in the world at 1.5 quadrillion cubic feet," he added. *Haynesville will become the largest US gas field by 2020,* he added. [Emphasis added][17]

As I discussed in Part I, more recent research on the Haynesville suggests that the play is unlikely to live up to its early hype. In late 2011, Louisiana State University researchers Mark J. Kaiser and Yunke Yu published a report titled "Louisiana Haynesville Shale—1: Characteristics, Production Potential of Haynesville Shale Wells Described," which made two important observations:

1. The average Haynesville well experiences a rapid production decline in its first two years online and can be expected to produce 80 percent of its expected recovery during this period.[18]
2. With regards to estimated recoveries the researchers found that "About 40% of Haynesville wells are expected to yield EURs between 2 and 3 bcf/well, 30% less than 2 bcf/well, and 30% greater than 3 bcf/well."[19]

Should the LSU researchers be correct in that the average Haynesville well will produce 80 percent of its lifetime total in its first two years and ultimately recover less than 3 bcf, the Haynesville will produce only a fraction of its early estimates. As I discussed earlier, Art Berman independently came to largely the same conclusion—that the average Haynesville well will likely produce approximately 3 bcf over its productive lifetime.

Current State of the Hayneville

The weak gas price environment of 2009 to 2012, significant financial distress by several of its leading operators and the end of drilling to hold leases in the play combined to greatly reduce activity by the summer of 2012. As you can see in Figure 11.05, by August 2012 there were only 26 wells actively drilling, down from 186 a year earlier.

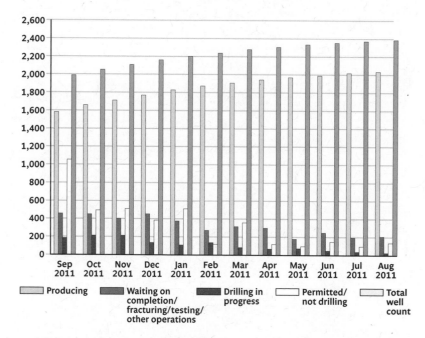

	Month	Producing	Waiting on completion/ fracturing/testing/ other operations	Drilling in progress	Permitted/ not drilling	Total well count
1	Sep 2011	1,615	467	186	1,074	2,030
2	Oct 2011	1,691	464	214	500	2,093
3	Nov 2011	1,752	410	214	525	2,150
4	Dec 2011	1,811	460	140	390	2,202
5	Jan 2012	1,870	388	112	525	2,251
6	Feb 2012	1,910	285	140	122	2,286
7	Mar 2012	1,942	323	87	366	2,320
8	Apr 2012	1,977	304	69	122	2,349
9	May 2012	2,010	182	77	98	2,373
10	Jun 2012	2,036	258	46	146	2,397
11	Jul 2012	2,055	198	36	98	2,415
12	Aug 2012	2,071	205	26	134	2,432

FIGURE 11.05. Haynesville Shale Well Activities by Month.[20]

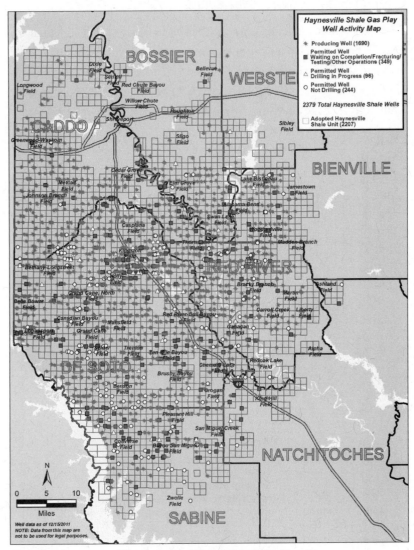

FIGURE 11.06. Haynesville Shale Gas Play Well Activity Map.[21]

The 86 percent drop in drilling over the course of 12 months is possibly the best indicator of the future deliverability of the play. After peaking at an estimated 6 bcf/d (the Louisiana Department of Natural Resources does not separate Haynesville production from its statewide figures) in late 2011, production from the Haynesville is declining.

Lastly, when attempting to determine the ultimate recovery of any

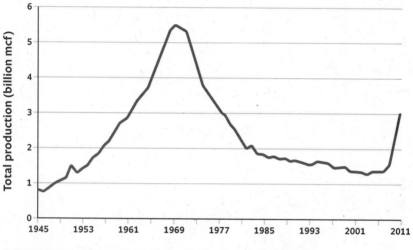

FIGURE 11.07. Louisiana Natural Gas Production 1945–2011.[22]

play it is important to understand its aerial extent, which limits the number of potential drilling locations. Unless successful Haynesville wells can be drilled outside of the areas currently under development, the ultimate recovery of the play will be only a fraction of early estimates. Figure 11.06, taken from the Louisiana DNR website in September 2012, shows how tightly clustered Haynesville wells are around the parishes of Caddo, DeSoto, Red River, Bienville, Sabine and Bossier.

Statewide Production History

While production growth in the Haynesville had a profound impact on the state's and the nation's production between 2008 and 2011, gas production in Louisiana seems to have resumed its multi-decade slide. As you can see from Figure 11.07, prior to the development of the Haynesville, the state's production had entered a terminal decline, which I believe has recently resumed.

CHAPTER 12

The Gulf of Mexico's Terminal Decline

W HILE THE FEDERAL waters of the Gulf of Mexico (GOM) have been a major source of natural gas supply for the US for decades, the area, which as of this writing in late 2012 accounts for approximately 6 percent of US production, has entered a terminal decline that shows no sign of abating.[1] Gas production from the GOM federal waters, which begin three miles from shore, is declining because of all of the same challenges facing Louisiana's offshore gas industry, which I discussed in Chapter 11. Operators in federal waters face the added costs associated with working in deeper water.

Though new technology may allow for the development of potentially large and previously untouchable natural gas and oil reserves below the salt weld throughout the GOM, the advanced stage of maturity of the GOM is often overlooked. The inability to replace annual production with new reserves is the hallmark of a very mature petroleum-producing region, and this is exactly the scenario that has been unfolding in the GOM for years. Through the end of 2006, the federal waters of the GOM had produced an incredible 15 billion barrels of oil and 167 tcf of natural gas from more than 1,200 fields.[2] Figure 12.01 shows that an estimated three-quarters of the total natural gas endowment of the GOM natural gas has been produced through the end of 2006.

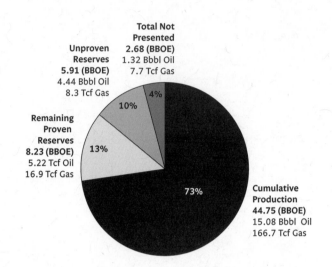

FIGURE 12.01. Gulf of Mexico Reserves and Resources.[3]

According to a December 2009 assessment of the federal GOM by the US Bureau of Ocean Energy Management, Regulation and Enforcement (formerly known as the Minerals Management Service), as of December 31, 2006, the federal waters of the GOM have remaining proven gas reserves of 16.9 tcf and an additional 8.3 tcf of unproven reserves.[4] Proven and unproven reserves together total 25.2 tcf, approximately enough to supply the US with natural gas for one year or about 12 years of production at the 2011 annual production rate of approximately 5 bcf/d.

Table 12.01 shows that remaining GOM proven natural gas reserves have been declining since 1985, when they reached a peak of 45.8 tcf.

More importantly, if we look at current reserve estimates from a historical perspective, it is quite clear that the best days of natural gas production from the GOM have passed. According to the above-cited 2009 report by the MMS, *the 10 largest gas fields ever tapped in the GOM were discovered before 1964.* Though MMS distinguishes between oil and gas fields, fields that are designated as oil fields also often produce substantial amounts of natural gas. Table 12.02 provides an excellent summary of the history of hydrocarbon discovery and production in the GOM. Natural gas fields are denoted with a "G" in the "Field type" column.

TABLE 12.01. Proven oil and gas reserves and cumulative production at end of year, 1975–2006 for Gulf of Mexico, Outer Continental Shelf and Slope.[5]

Year	Number of fields included	Proven reserves			Historical cumulative production			Remaining proven reserves		
		Oil	Gas	BOE	Oil	Gas	BOE	Oil	Gas	BOE
1975	255	6.61	59.9	17.27	3.82	27.2	8.66	2.79	32.7	8.61
1976	306	6.86	65.5	18.51	4.12	30.8	9.60	2.74	34.7	8.91
1977	334	7.18	69.2	19.49	4.47	35.0	10.70	2.71	34.2	8.80
1978	385	7.52	76.2	21.08	4.76	39.0	11.70	2.76	37.2	9.38
1979[1]	417	7.71	82.2	22.34	4.83	44.2	12.69	2.88	38.0	9.64
1980	435	8.04	88.9	23.86	4.99	48.7	13.66	3.05	40.2	10.20
1981	461	8.17	93.4	24.79	5.27	53.6	14.81	2.90	39.8	9.98
1982	484	8.56	98.1	26.02	5.58	58.3	15.95	2.98	39.8	10.06
1983	521	9.31	106.2	28.21	5.90	62.5	17.02	3.41	43.7	11.19
1984	551	9.91	111.6	29.77	6.24	67.1	18.18	3.67	44.5	11.59
1985	575	10.63	116.7	31.40	6.58	71.1	19.23	4.05	45.6	12.16
1986	645	10.81	121.0	32.34	6.93	75.2	20.31	3.88	45.8	12.03
1987	704	10.76	122.1	32.49	7.26	79.7	21.44	3.50	42.4	11.04
1988[2]	678	10.95	126.7	33.49	7.56	84.3	22.56	3.39	42.4	10.93
1989	739	10.87	129.1	33.84	7.84	88.9	23.66	3.03	40.2	10.18
1990	782	10.64	129.9	33.75	8.11	93.8	24.80	2.53	36.1	8.95
1991	819	10.74	130.5	33.96	8.41	98.5	25.94	2.33	32.0	8.02
1992	835	11.08	132.7	34.69	8.71	103.2	27.07	2.37	29.5	7.62
1993	849	11.15	136.8	35.49	9.01	107.7	28.17	2.14	29.1	7.32
1994	876	11.86	141.9	37.11	9.34	112.6	29.38	2.52	29.3	7.73
1995	899	12.01	144.9	37.79	9.68	117.4	30.57	2.33	27.5	7.22
1996	920	12.79	151.9	39.82	10.05	122.5	31.85	2.74	29.4	7.97
1997	957	13.67	158.4	41.86	10.46	127.6	33.17	3.21	30.8	8.69
1998	984	14.27	162.7	43.22	10.91	132.7	34.52	3.36	30.0	8.70
1999	1,003	14.38	161.3	43.08	11.40	137.7	35.90	2.98	23.6	7.18
2000	1,050	14.93	167.3	44.70	11.93	142.7	37.32	3.00	24.6	7.38
2001	1,086	16.51	172.0	47.11	12.48	147.7	38.77	4.03	24.3	8.35
2002	1,112	18.75	176.8	50.21	13.05	152.3	40.15	5.71	24.6	10.09
2003	1,141	18.48	178.2	50.19	13.61	156.7	41.49	4.87	21.5	8.70
2004	1,172	18.96	178.4	50.70	14.14	160.7	42.73	4.82	17.7	7.97
2005	1,196	19.80	181.8	52.15	14.61	163.9	43.77	5.19	17.9	8.38
2006	1,229	20.30	183.6	52.97	15.08	166.7	44.74	5.22	16.9	8.23

[1] Gas plant liquids dropped from system.
[2] Basis of reserves changed from demonstrated to SPE proved.

It should not come as a surpise to serious students of the oil and gas industry that the largest fields in the GOM were the first to be discovered. A similar pattern of discovery occurs in nearly every hydrocarbon-producing region. For example, the super-giant fields of the Middle East such as Ghawar in Saudi Arabia or Kuwait's Burgan complex were among the first to be found in the region. More importantly, these

TABLE 12.02. Gulf of Mexico proven fields by rank order, based on probed barrels of oil equivalent reserves, top 50 fields.[6]

Rank	Field name	New field	Disc year	Water depth (ft.)	Field class	Field type	Field GOR (SCF/STB)	Proven reserves			Cumulative production through 2006			Remaining proven reserves		
								Oil (MMbbl)	Gas (Bcf)	BOE (MMbbl)	Oil (MMbbl)	Gas (Bcf)	BOE (MMbbl)	Oil (MMbbl)	Gas (Bcf)	BOE (MMbbl)
1	MC807		1989	3,393	PDP	O	1,444	1,208.2	1,745.2	1,518.7	734.7	959.6	905.4	473.5	785.6	613.3
2	EI330		1971	247	PDP	O	4,222	430.9	1,819.5	754.7	420.3	1,801.4	740.9	10.6	18.1	13.8
3	WD030		1949	48	PDP	O	1,617	573.7	927.7	738.7	561.8	867.8	716.2	11.9	59.9	22.6
4	MC778		1999	6,081	PU	O	776	642.7	498.4	731.4	0.0	0.0	0.0	642.6	498.4	731.0
5	GI043		1956	140	PDP	O	4,302	377.3	1,618.9	665.3	360.8	1,537.1	634.3	16.5	81.8	31.0
6	MC776	*	2000	5,662	PU	O	1,058	534.0	565.2	634.5	0.0	0.0	0.0	534.0	565.2	634.5
7	BM002		1949	50	PDP	O	1,037	530.3	549.9	628.1	522.5	536.5	618.0	7.7	13.4	10.1
8	GC743	*	1998	6,468	PDP	O	647	558.6	361.4	623.0	0.0	0.0	0.0	558.6	361.4	622.9
9	TS000		1958	13	PDP	G	83,526	38.3	3,201.4	608.0	37.5	3,155.0	598.8	0.9	46.4	9.1
10	VR014		1956	26	PDP	G	63,983	48.2	3,082.6	596.7	47.9	3,055.7	591.6	0.3	26.8	5.1
11	MP041		1956	42	PDP	O	5,715	263.0	1,503.1	530.5	252.1	1,448.2	509.8	10.9	55.0	20.7
12	VR039		1948	38	PDP	G	81,151	31.7	2,572.6	489.5	31.2	2,542.9	483.6	0.5	29.7	5.8
13	SS208		1960	102	PDP	O	6,217	220.3	1,369.5	464.0	216.0	1,338.5	454.2	4.3	30.9	9.8
14	GC640	*	2002	4,234	PDN	O	487	414.0	201.6	449.9	0.0	0.0	0.0	414.0	201.6	449.9
15	WD073		1962	178	PDP	O	2,458	265.2	651.7	381.1	259.3	632.0	371.7	5.9	19.7	9.4
16	GB426		1987	2,860	PDP	O	3,579	229.0	819.4	374.8	211.7	757.9	346.5	17.3	61.5	28.2
17	GI016		1948	53	PDP	O	1,271	303.4	385.5	372.0	299.2	377.9	366.4	4.2	7.6	5.6
18	SP061		1967	219	PDP	O	1,930	266.9	515.1	358.5	259.5	505.1	349.4	7.4	10.0	9.1
19	ST021		1957	46	PDP	O	1,729	272.7	471.5	356.6	246.0	396.5	316.6	26.7	74.9	40.0
20	EI238		1964	147	PDP	G	16,327	91.2	1,489.5	356.3	85.8	1,423.9	339.1	5.4	65.6	17.1
21	ST172		1962	98	PDP	G	136,478	14.0	1,907.2	353.3	11.5	1,831.9	337.4	2.5	75.4	15.9
22	SP089		1969	423	PDP	O	4,448	191.1	849.9	342.3	188.3	826.4	335.3	2.8	23.5	7.0
23	WC180		1961	48	PDP	G	141,655	12.9	1,821.4	336.9	12.7	1,779.5	329.3	0.2	41.9	7.6
24	AC857	*	2002	7,900	PU	O	1,205	272.5	328.3	330.9	0.0	0.0	0.0	272.5	328.3	330.9
25	ST176		1963	126	PDP	G	14,710	89.7	1,320.0	324.6	81.5	1,171.5	290.0	8.2	148.5	34.7
26	SS169		1960	63	PDP	O	5,411	163.2	883.3	320.4	154.3	825.1	301.2	8.9	58.1	19.2

27	SM048	1961	101	PDP	G	55,963	28.6	1,601.1	313.5	27.8	1,512.7	297.0	0.8	88.4	16.5
28	MC194	1975	1,022	PDP	O	4,175	178.8	746.4	311.6	176.5	738.0	307.8	2.3	8.4	3.8
29	EC064	1957	50	PDP	G	57,810	27.4	1,586.2	309.7	26.6	1,537.9	300.3	0.8	48.4	9.4
30	EI292	1964	212	PDP	G	84,604	19.1	1,617.4	306.9	18.3	1,609.4	304.7	0.8	7.9	2.2
31	EC271	1971	171	PDP	G	18,853	70.3	1,325.8	306.2	67.5	1,309.3	300.5	2.8	16.5	5.7
32	SS176	1956	100	PDP	G	19,836	65.3	1,294.6	295.6	62.9	1,261.7	287.4	2.3	32.9	8.2
33	SP027	1954	64	PDP	O	5,219	151.7	791.6	292.5	150.0	762.3	285.7	1.7	29.3	6.9
34	WC587	1971	211	PDP	G	110,142	14.1	1,554.0	290.6	12.8	1,528.5	284.8	1.3	25.5	5.8
35	ST135	1956	130	PDP	O	3,612	171.7	620.0	282.0	165.7	579.5	268.8	6.0	40.6	13.2
36	EI296	1971	214	PDP	G	69,965	20.3	1,421.6	273.3	20.3	1,413.6	271.8	0.0	8.0	1.5
37	WC192	1954	57	PDP	G	58,762	23.8	1,399.6	272.9	22.3	1,356.8	263.7	1.5	42.8	9.1
38	WD079	1966	124	PDP	O	3,800	162.7	618.3	272.7	160.5	609.1	268.9	2.2	9.2	3.8
39	MI623	1980	83	PDP	G	98,785	14.4	1,426.2	268.2	13.3	1,335.0	250.9	1.1	91.2	17.3
40	HI573A	1973	341	PDP	O	7,700	111.2	856.2	263.5	107.6	850.1	258.9	3.6	6.1	4.6
41	GC644	1999	4,340	PDP	O	1,234	209.6	258.7	255.6	28.0	29.4	33.3	181.5	229.3	222.3
42	GI047	1955	88	PDP	O	3,583	150.1	538.0	245.8	144.2	516.2	236.1	5.9	21.7	9.8
43	SP078	1972	203	PDP	G	11,544	77.6	896.3	237.1	72.9	881.3	229.8	4.7	15.0	7.4
44	SM023	1960	82	PDP	G	38,903	29.7	1,155.4	235.3	29.5	1,143.8	233.0	0.2	11.7	2.3
45	SM130	1973	214	PDP	O	1,341	187.4	251.3	232.1	182.8	246.0	226.6	4.5	5.3	5.5
46	PL020	1951	33	PDP	O	5,810	113.7	660.3	231.2	108.1	604.8	215.7	5.5	55.5	15.4
47	GC244	1994	2,762	PDP	O	2,005	170.3	341.5	231.0	160.0	318.8	216.7	10.3	22.7	14.3
48	VR076	1949	31	PDP	G	140,837	8.7	1,231.9	228.0	7.4	1,168.8	215.4	1.4	63.1	12.6
49	SM066	1963	124	PDP	G	255,946	4.9	1,250.3	227.4	4.8	1,218.0	221.5	0.1	32.3	5.9
50	VK956	1985	3,254	PDP	O	9,042	87.1	787.3	227.2	80.2	710.8	206.7	6.9	76.5	20.5

Middle Eastern discoveries, as well as hundreds of others in the region, were made using exploration techniques that would be considered primitive by today's standards. Explorers simply looked for topographical features that were known indicators of hydrocarbons such as anticlines as well as areas where oil and gas historically seeped to the surface (think burning bush) to identify some of the world's largest deposits of oil and gas. This pattern has repeated itself in the Gulf of Mexico. No need for seismic ships equipped with the latest seismic bouys and supercomputers—the majority of GOM reserves were discovered decades ago using far less sophisticated technology.

One of the best indicators of future production is the number of rigs drilling for natural gas. The maturity of the GOM, combined with several damaging hurricanes in the past decade and the April 2010 Deepwater Horizon (Macondo) tragedy and subsequent oil spill, have greatly

TABLE 12.03: GOM Natural Gas Rig Count, Daily Production and Prices 1992 thru 2011.[7]

Year	Average NG rigs in GOM	Avg. daily marketed GOM NG production in bcf/d	Average wellhead NG price
1992	NA	12.81	$1.74
1993	NA	12.83	$2.04
1994	61	13.45	$1.85
1995	63	13.30	$1.55
1996	91	14.12	$2.17
1997	99	14.26	$2.32
1998	91	13.91	$1.96
1999	80	13.78	$2.19
2000	117	13.52	$3.68
2001	118	13.77	$4.00
2002	95	12.36	$2.95
2003	98	12.07	$4.88
2004	90	10.88	$5.46
2005	75	8.58	$7.33
2006	81	7.95	$6.39
2007	68	7.67	$6.25
2008	61	6.34	$7.96
2009	38	6.65	$3.71
2010	16	6.15	$4.48
2011	18	4.98	$3.95

Source: Baker Hughes, EIA.

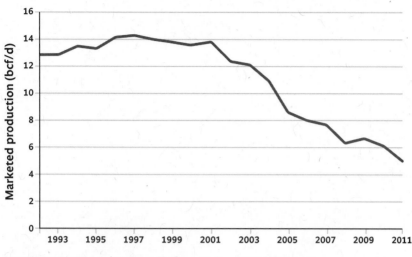

FIGURE 12.02. Natural Gas Supply from Federal GOM Waters.[8]

reduced natural gas-directed drilling in the Gulf. Table 12.03 shows that such drilling activity in the GOM has remained near historic lows for much of the past three years.

The advanced maturity of the GOM becomes quite clear when looking at the relationship between drilling, production and prices in Table 12.03. The lack of a production response to higher prices after the turn of the millenium is astonishing considering the thousands of miles of gathering infrastructure already in place along the seabed floor that could have been used to enhance the economics of any successful exploration efforts. Declining production in the face of rising prices is a sure sign that a region is in terminal decline.

The federal waters of the Gulf of Mexico, once the backbone of America's natural gas industry, are in their twilight as a significant natural gas-producing region. Despite being the testing ground for nearly every major advancement in offshore technology over the past sixty years, the GOM has seen its natural gas production decline since peaking in 1997. While there exists some hope that the potential riches below the salt weld may someday reverse the GOM's decline, as of late 2012, no gas is being produced from below the weld and it will likely be years (if ever) before ultra-deep wells in the GOM make a meaningful contribution to America's gas supply.

CHAPTER 13

The Cowboy State's Reversal of Fortune

D UE TO THE advances in multi-zone fracturing techniques, growing CBM production and the build-out of Rocky Mountain pipeline capacity, Wyoming experienced a large ramp in gas production
between 1995 and 2009. State production grew from 1.84 bcf/d in 1995 to
6.4 bcf/d in 2009, a 248 percent increase over 14 years.[1] However, in line
with the financial world's well-worn caveat that "past performance may
not be representative of future performance," Wyoming recently moved
from a state that was rapidly growing natural gas production to one that
will likely experience a large drop over the next decade. In fact, since the
2009 peak, production had fallen by more than 10 percent by mid-2012
and the rate of decline looks to be accelerating.[2] In this chapter I will examine the history of natural gas production in Wyoming as well as take
an in-depth look at several of the state's largest fields, which comprise
approximately 71 percent of its production.[3]

Pinedale Field

Located in Wyoming's Green River Basin and straddling the Pinedale
anticline, the Pinedale field covers approximately 90 square miles of
Sublette County.[4] Operators began drilling on the field in the late 1930s
and had little success at first, due to the tightness of the gas-bearing

sandstones they were targeting.[5] However, after the engineers at McMurray Oil Company were able to use modern fracturing techniques to coax gas out of the tight sands in the nearby Jonah field in the early 1990s, operators soon discovered the same approach could be used at Pinedale.[6] Since commercial development got into full swing in the Pinedale in 1997, more than 1,400 wells have been drilled and 3.36 tcf have been produced through year-end 2011.[7] In 2011, the Pinedale field produced an average of 1.52 bcf/d,[8] making it the state's largest producing field. While production has been on a plateau since 2010, substantial resources remain in the field, which should allow production to increase once gas prices rebound. Figure 13.01 is a map displaying both the Pinedale and Jonah fields.

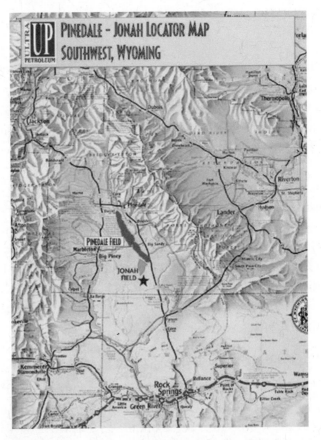

FIGURE 13.01. Pinedale and Jonah Field Locator Map.[9]

Jonah Field

Wyoming's Jonah field, the state's first significant tight sands field to come under commercial development, has been a prolific producer in its first two decades. According to the Wyoming Oil and Gas Conservation Commission, the field produced approximately 3.58 tcf of natural gas between 1992 and 2011.[10] Located just south of the Pinedale field, the Jonah field covers 36 square miles; like the Pinedale, its two main producing horizons are the Lance and Mesaverde formations.[11] After peaking in 2008 at 1.12 bcf/d, Jonah production dropped to 0.84 bcf/d in 2011 and appears to have entered terminal decline.[12] Given that 1,500 wells, or about one well per 40 acres, have already been drilled into it, the field is nearly fully developed.[13]

Powder River Basin CBM

Wyoming's Powder River Basin (PRB) coal bed methane (CBM) play is the world's second largest CBM field and accounts for nearly all CBM produced in the state. Since the late 1990s—when Section 29 credits were still available—more than 26,000 CBM wells have been drilled in the PRB.[15] The ability of operators to put together large tracts of prospective and contiguous land combined with the shallow depth and high permeability of the basin's two most prolific coal seams, the Wyodak and the Big George, encouraged rapid development. As a result of intense development efforts production grew rapidly for a decade before peaking in 2009 at 544 bcf per annum.[16] Figure 13.03 shows Wyoming CBM production thru mid-2012.

While the PRB CBM play has produced 4.73 tcf since commercial development of the play began in 1997, the play has also generated controversy over the water produced as part of the CBM extraction process.[18] For example, the average CBM well in the PRB produces 150 barrels of water a day (bpd), but in initial stages, water production can range from 400 to 1,500 bpd. For comparision purposes, wells in the San Juan Basin CBM field produce approximately 25 bpd of water, the vast majority of which is reinjected into deep formations.[19] However, the PRB contains few water disposal wells, so the majority of water produced along with the CBM is contained in evaporation pits (64 percent) or discharged

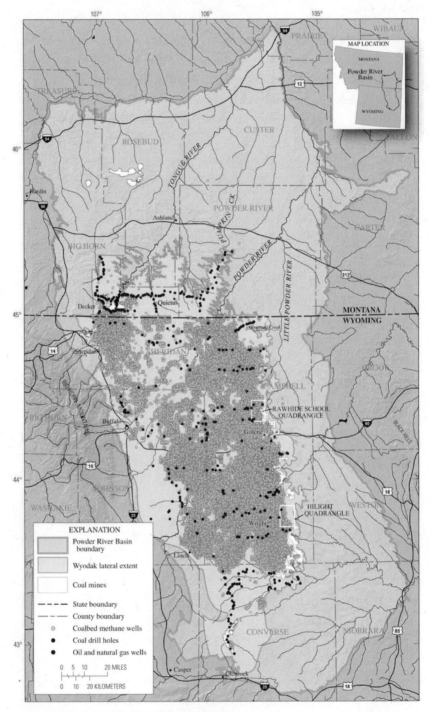

FIGURE 13.02. Powder River Basin CBM.[14]

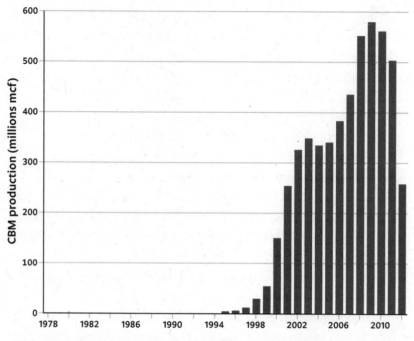

FIGURE 13.03. Wyoming Coalbed Methane Production (mcf).[17]

into streams (20 percent);[20] another 13 percent is used for irrigation.[21] While some surface owners have used produced water for cattle or crop irrigation, others have objected to the building of containment ponds and the discharge of produced water into streams.

Sour Gas In Wyoming

In addition to Wyoming's substantial tight sands and CBM production, the state is also home to two significant sour gas fields. Sour gas is generally defined as natural gas that contains enough hydrogen sulfide that it requires processing before it can be sold into the pipeline system. Covering approximately 63 square miles, the LaBarge Complex in Sublette County is the state's largest sour gas-producing area (it is actually three fields). LaBarge is operated by ExxonMobil,[22] which processes all raw LaBarge gas, approximately 720 mmcf/d, at its Shute Creek and Black Canyon gas plants.[23] The average composition of the natural gas produced at LaBarge is approximately 66% carbon dioxide, 22% methane, 4% hydrogen sulfide, 7.5% nitrogen and 0.5% helium.[24]

The carbon dioxide is used in enhanced oil recovery (EOR) at several oil fields in Wyoming as well as one oil field in Colorado. According to a representative at the Wyoming State Geological Survey, "the methane is sold as fuel; the helium production is over 1 bcf per year which makes [LaBarge] the largest helium-producing area in the US."[25] Due to facility constraints and the large size of the sour natural gas fields at LaBarge, production will likely remain at 720 mmcf/d for decades.

Another source of sour gas in Wyoming is the deep (up to 25,000 feet) Madison Limestone of the Madden Field in the Wind River Basin.[26] Wells completed in the deep Madison have very high deliverability of up to 45 mmcf/d and produce raw gas that contains approximately 12 percent hydrogen sulfide (H_2S), 19 percent carbon dioxide (CO_2) and 69 percent methane.[27] All gas from the Madden field—315 mmcf/d—is processed at the ConocoPhillips-owned Lost Cabin gas plant.[28] Denbury Resources (NYSE:DNR) purchases CO_2 from Lost Cabin and ships it for reinjection into its Belle Creek EOR project in Montana via a recently constructed 232-mile pipeline.[29] Denbury estimates that Belle Creek may contain up to 200 million barrels of recoverable oil and plans to initiate its CO_2 flood at Belle Creek in 2013.[30]

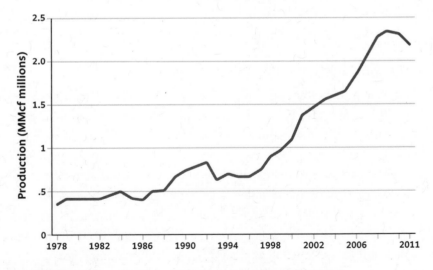

FIGURE 13.04. Wyoming State Wide Gas Production 1978–2011.[31]

Wyoming Statewide Production

While the growth in gas production in Wyoming over the past two decades has been nothing short of remarkable, all good things must come to an end. While the state's largest field, the Pinedale, will likely be able to keep production stable for several more years, the Jonah and PRB CBM fields have undoubtedly entered terminal decline. Though there are large reserves in the Wind River Basin and the LaBarge Complex that will be a major source of gas for decades to come, the need for expensive processing equipment will likely keep sour gas production from expanding materially. Due to the advanced age of Wyoming's major sources of production and the lack of significant new discoveries in recent years, the state's production has entered a terminal decline.

Figure 13.04 is a graphic taken from the Wyoming Oil and Gas Conservation Commission website showing the growth in gas production from 1978 thru 2011.

CHAPTER 14

New Mexico's
Long Slide

THOUGH THE Land of Enchantment (New Mexico's motto) has
fallen off the radar of most energy analysts since it is not home to
any large shale plays, the state is still a very important supplier that produces approximately 6 percent of America's total gas production and is
home to the world's largest coal bed methane (CBM) field.[1]

As you can see from Figure 14.01 displaying New Mexico's major
sources of hydrocarbons, the state has two areas of significant gas production, the San Juan and Permian basins.

New Mexico began producing oil and gas in the 1920s and became
ground zero for the unconventional gas industry in the 1980s. In 1977
Amoco Production Company discovered the world's first significant
CBM field in New Mexico's San Juan Basin, and shortly thereafter began
a pilot program to test its new discovery.[3] While initial results in the
very high-quality Fruitland coals were positive, it took the passage of the
Section 29 credits in 1980 to bring other operators to the Basin. By the
mid-1980s more than thirty operators, including some of America's largest independents, had started pilot programs in the play.[4] The Colorado
portion of the San Juan CBM play would be developed in later years.
The combination of Section 29 credits, existing pipeline capacity (due
to legacy conventional gas production), low-cost water disposal options,

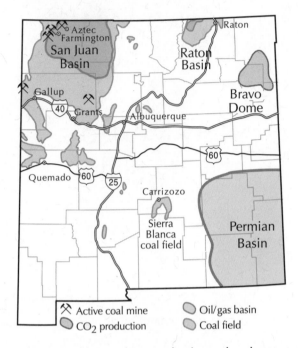

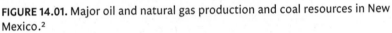

FIGURE 14.01. Major oil and natural gas production and coal resources in New Mexico.[2]

large reserves and the high deliverability of Fruitland CBM wells encouraged operators to ramp up production despite the weak gas prices of the late 1980s.

The Fruitland coal seams of the San Juan Basin are found 600 to 3,500 feet below the surface and wells producing from them have average peak production of between 700 thousand cubic feet and 1 million cubic feet of gas per day (mmcf/d).[5] A typical well can be expected to recover 2 billion cubic feet (bcf) over its lifetime, far more than any other CBM play in the US.[6] Through the end of 2010, the San Juan Basin (including the Colorado portion) has produced approximately 16 tcf of CBM and has accounted for approximately 66 percent of all CBM ever produced in the US.[7] Figure 14.02 not only illustrates the large ramp-up in CBM production in the San Juan Basin during the 1990s, but also puts into perspective the field's size relative to other CBM basins in the US (note that the graphic includes production from the Colorado portion of the San Juan Basin).

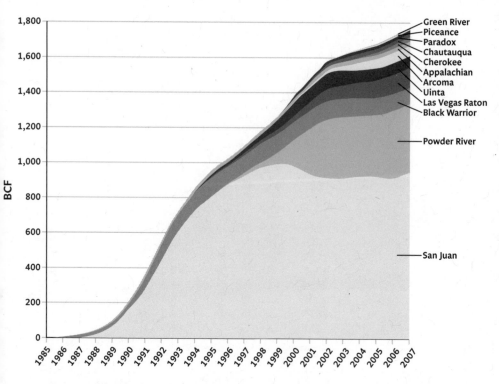

FIGURE 14.02. CBM Production History by Basin in the US.[8]

CBM production in New Mexico's portion of the San Juan Basin peaked in 1999 at 612 bcf and remained fairly stable for eight more years as operators down-spaced the field and optimized existing wells.[9] However, as shown in Table 14.01, CBM production has fallen more rapidly in recent years and is now more than 42 percent below its 1999 peak. The table contains data provided by the New Mexico Oil Conservation Division for San Juan CBM production from 1994 through 2011.

In a presentation he gave in June 2010, US Geological Survey Scientist Emeritus James E. Fassett provided many interesting facts about the San Juan Basin, including the following (once again these figures include the Colorado portion of the Basin):[11]

- The SJ is the second largest gas basin in the US behind the Hugoton Complex of Kansas, Oklahoma and Texas.
- Conventional fractured sandstone reservoirs have produced 24.4 tcf of gas through September 2009.

TABLE 14.01. New Mexico Coalbed
Methane Gas Production.[10]

Year	NM San Juan CBM gas in bcf	Percentage change
1994	531	NA
1995	572	+7.7%
1996	603	+5.4%
1997	601	0.0%
1998	608	+1.1%
1999	612	+.6%
2000	584	−4.5%
2001	533	−8.7%
2002	486	−8.1%
2003	464	−4.5%
2004	485	+4.5%
2005	494	+1.8%
2006	498	+.8%
2007	482	−3.2%
2008	450	−6.6%
2009	424	−5.7%
2010	381	−10.1%
2011	353	−7.3%

Source: New Mexico Oil Conservation Division.

- The Basin contains more than 150 gas fields and more than 39,000 wells.
- Field-down spacing has greatly enhanced conventional gas production.

Figure 14.03 shows how densely drilled the SJ Basin has become.

In addition to the San Juan Basin, New Mexico is home to a portion of the Las Vegas Raton CBM play. Similar to the San Juan play, the Raton Basin straddles the New Mexico–Colorado border (see Figure 14.01). In the mid-1990s several operators, led by play leader Evergreen Resources, began producing small amounts of CBM from the Vermejo Formation coal bed in Colfax County.[13] As gas prices rose significantly after the turn of the millennium, so did CBM production from the Raton Basin, from slightly more than 1 bcf in 2000 to a record 26.5 bcf in 2011.[14]

The final major area of natural gas production in New Mexico is the state's share of the prolific Permian Basin, one of America's largest hydrocarbon-producing basins. Despite a pick-up in drilling activity in

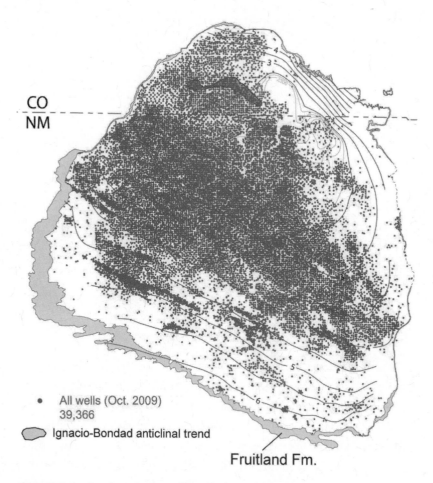

CO
NM

- All wells (Oct. 2009)
  39,366

Ignacio-Bondad anticlinal trend

Fruitland Fm.

FIGURE 14.03. San Juan Basin Wells Drilled—October 2009.[12]

recent years as operators pursue the oil-prone Avalon and Wolfcamp Shale formations in Lea and Eddy counties, natural gas production in the New Mexico portion of the Permian remains in terminal decline. In the decade between 2002 and 2011, it dropped from 587 bcf per annum to only 411 bcf per annum, a decline of 30 percent.

Due to the advanced maturity of New Mexico's two major gas-producing basins, the San Juan and Permian basins, annual natural gas production in New Mexico has been on a steady downward path since peaking in 2001. Figure 14.04 displays gas production in New Mexico for years 1970 to 2011.

FIGURE 14.04. New Mexico Natural Gas Production 1970–2011.[15]

The production decline of more than 400 bcf per annum, approximately 1.1 bcf/d or 24 percent, over the course of the decade between 2002 and 2011, is proof positive that New Mexico gas production has entered a terminal decline. With the discovery cupboard now nearly bare and many of the state's largest fields nearly completely drilled out, there is little that can be done to slow down New Mexico's dropping natural gas production.

CHAPTER 15

Pennsylvania's
Marcellus Renaissance

THE COMMERCIALIZATION of the Marcellus Shale has been a godsend to Pennsylvania's oil and gas industry. The state had seen numerous years of flat gas production prior to the development of the Marcellus, with few new discoveries to replace depleting reserves. Now, everything has changed. In just a few short years, Pennsylvania witnessed a voracious land grab followed by one of the biggest drilling booms of all time. However, the rapid development of the Marcellus has created numerous environmental issues for the people of Pennsylvania and has heightened the national debate over the safety of hydraulic fracturing. In this chapter I will briefly review Pennsylvania's long history of natural gas production with a focus on the Marcellus Shale and address some of the environmental concerns surrounding hydraulic fracturing.

Early Devonian Shale and Oriskany

Pennsylvania is widely recognized as the birthplace of America's oil industry. Colonel Drake drilled the country's first oil well near Oil Creek in 1859, and the state saw its first commercial natural gas production from shallow, organic-rich Devonian shale by the mid-1800s.[1] Shortly after America's first natural gas well was drilled in Fredonia, NY in 1821, the country's first natural gas drilling boom began along the shores of Lake Erie as residents of the area tapped into the shallow, naturally fractured Devonian shale to heat homes and power factories.[2] By the beginning of the twentieth century, nearly every home and factory within a mile of the shores of Lake Erie had its own natural gas well.[3] Many of the wells, typically less than a thousand feet deep, produced for decades, though pressure changes due to weather variations impacted their productivity at times. Some wells drilled in the early twentieth century still provide gas to residents of Erie County today.[4]

In 1930, the North Penn Gas Company drilled the first successful Oriskany sandstone well, in Tioga County.[5] The Oriskany sandstone, which is found directly below the Marcellus Shale throughout much of the state, has now been a steady producer for over eighty years, with more than 1,700 wells drilled to date.[6] The formation continues to produce gas and is used as gas storage in north-central Pennsylvania. Figure 15.01 is a map of Oriskany fields.

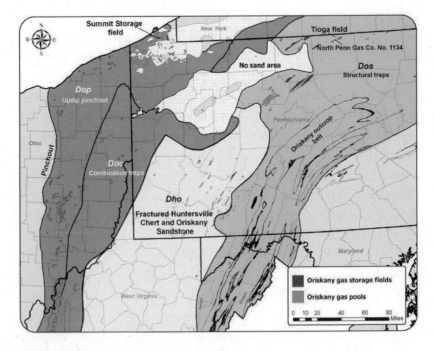

FIGURE 15.01. Map of the Oriskany Play.[7]

The Marcellus Changes Pennsylvania Forever

While the Oriskany played an important role in the early growth of the state's gas industry, it also provided today's Marcellus developers with significant insight. When drilling Oriskany wells, many operators noticed significant gas kicks while drilling through the Marcellus Shale, and drilling often had to be halted to let the gas dissipate. Further evidence of the potential of the Marcellus was confirmed when the Department of Energy undertook the Eastern Shale Gas Project (EGSP) in the 1970s and '80s. One of the most important studies undertaken during the EGSP was the mapping of the Marcellus and other Devonian-aged shales through the use of gamma ray logs.[8] Figure 15.02 is a map showing the varying thickness of the Marcellus and other shales throughout Pennsylvania.

It should be noted that nearly all development of the Marcellus in Pennsylvania as of mid-2012 is within the area the EGSP rated as being greater than one hundred feet thick.

Though the EGSP identified the Marcellus as a promising shale resource as far back as the 1970s, it remained undeveloped because of the

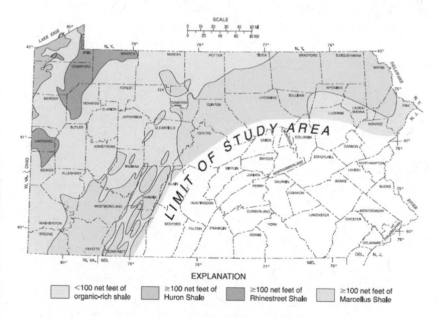

FIGURE 15.02. Distribution of Thickest Sequences of Shale.[9]

lack of available extraction technology and the low gas prices that persisted until shortly after the turn of the millennium. It wasn't until 2004, when Range Resources (NYSE:RRC), an early operator in the Barnett Shale, drilled the Renz #1 well in southwestern Pennsylvania, that commercial development of the Marcellus got underway.[10] Range and other early operators had a very good indication of the most prospective areas of the Marcellus through their drilling in the deeper Oriskany and Trenton Black River formations (many of the early Marcellus operators originally acquired their acreage in Pennsylvania to pursue the Trenton Black River formation and later abandoned the play due to high costs and inconsistent results). By 2006, dozens of operators had leased acreage throughout the still largely unproven Marcellus fairway. Land prices skyrocketed from the typical $25 per acre with a 12.5 percent royalty in the early days to over $10,000 per acre with a 25 percent royalty by early 2012.[11] Larger companies that were not able to lease meaningful acreage positions acquired companies that had been early participants in the Marcellus land rush.[12]

As in other modern shale plays that have been commercialized in the US, such as the Barnett, Fayetteville and Haynesville, operators in the Marcellus aggressively ramped up production to hold leases. As you can

TABLE 15.01. Pennsylvania Annual Production 2000 thru mid-2012 in Billion Cubic Feet.[13]

Year	Conventional	Marcellus	Total
2000	NA	NA	150
2001	NA	NA	130
2002	NA	NA	157
2003	NA	NA	159
2004	NA	NA	197
2005	NA	NA	168
2006	NA	NA	175
2007	NA	NA	182
2008	NA	NA	185
2009	NA	NA	274
2010	NA	NA	573
2011	249	1,086	1,335
2012 (6 months)	NA	895	895

Source: EIA.

see from Table 15.01 below, during 2011 and 2012, Marcellus production grew at an incredible rate despite weak gas prices.

Unfortunately for operators in the Marcellus, the large ramp-up in drilling coincided with the low gas prices of 2009 through 2012, making much of the early development of the play uneconomic. However, this did not slow down the early Marcellus players who needed to drill to hold acreage and to show the investment community production growth. Drillers such as Range Resources, Cabot Oil and Gas (NYSE: COG) and Chesapeake Energy far outspent their cash flows to grow their production. By taking on debt, forming joint ventures with larger entities willing to overpay for access to the Marcellus, issuing additional shares and selling off assets, companies managed to fund their Marcellus drilling programs despite the weak gas prices.

Drilling to hold leases and grow production during a time of low natural gas prices had a profound impact on the financial soundness of the play's most aggressive operator, Chesapeake Energy. By mid-2012, the company found itself in the unenviable position of having to sell assets into a weak market and increase its debt load to $13 billion (not including its off-balance sheet volumetric production payment [VPP] obligations or its $3 billion in preferred stock) to fund its difficult transition to a more oil-focused company.[14] Chesapeake's years of outspending its cash from operations had finally caught up with it as it ran out of foreign companies willing to buy its shale gas assets or to enter into money-losing joint ventures. (See Chapter 5 for a discussion of CHK's business model of acreage flipping.) The company's financial position became so weak that it needed to take out a $4 billion bridge loan from a Goldman Sachs-led syndicate in May 2012 as it attempted to sell off billions of dollars of assets.[15]

What Does the Future Hold for the Marcellus?

As I discussed in Part One, there should be little doubt that the Marcellus is one of America's most important natural gas fields. The play had produced approximately 2.5 tcf by mid-2012 after only a few years of commercial development and there is every indication that there is much more gas in the ground. Despite the impressive production results to date, however, much remains unknown about the Marcellus. Similar to the Barnett, Fayetteville, Haynesville Shale and Arkoma Woodford, where the vast majority of production has come from the core areas of the plays, I expect drilling activity in the Marcellus to contract to its two core areas. Once the outer limits of these areas are established and additional production history is captured, the estimated ultimate recovery (EUR) of the play is likely to be far smaller than is currently estimated.

The Marcellus Becomes Ground Zero for Fracking Debate

Few topics in America have received more attention in recent years than the hydraulic fracturing (fracking) of shale gas wells. Hydraulic fracturing began in the late 1940s, though the practice of creating manmade fractures in rock surrounding wellbores to increase productivity has been around since the early years of the oil business. Early wells were stimulated with nitroglycerin, a very dangerous practice.[16] While hydraulically fracturing wells has been a common industry practice for six decades and an estimated one million wells have been hydraulically fractured in the US to date, concerns over its environmental impact have only arisen in the last decade. Figure 15.03 is a graphic of a multi-stage stimulation of a horizontal well; note that each of the fissures is a result of a separate frac stage.

When hydraulic fracturing began, virtually all wells were vertical and all fracture stimulations involved only one stage. Applying a multi-stage fracture stimulation to a horizontal well is a far more involved process. Today's hydraulic fracture stimulations can use several million gallons of water and require a fleet of a couple of dozen large, specialized trucks to pump water, sand and chemicals into microscopic cracks in shale rock in an effort to unlock previously trapped hydrocarbons.

A February 2012 study by PennEnvironment Research and Policy Center found that Marcellus development resulted in numerous environmental violations. The study, based on the Pennsylvania Department of Environmental Protection (DEP) violations' database, compiling all drilling-related violations for operators between January 1, 2008 and December 31, 2011, found that 3,355 violations were issued to 64 different

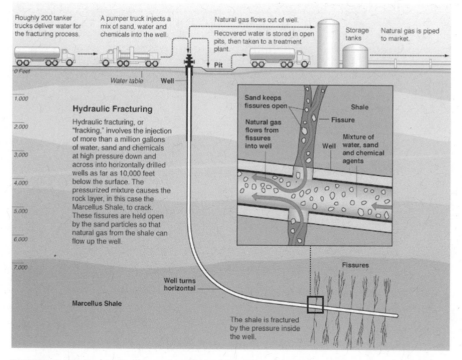

FIGURE 15.03. Hydraulic Fracturing Illustration.[17]

operators.[18] Figure 15.04, a graphic taken from the PennEnvironment study, neatly summarizes violations handed down by the DEP between 2008 and 2011.

While most of the violations were related to incidents that did not cause major damage to the environment, there were several notable exceptions. Here are a few examples:

- EOG Resources was fined more than $350,000 and threatened with the termination of its operator license in Pennsylvania following the June 2011 blow-out of a well in Clearfield County that spewed an estimated 35,000 gallons of fracture fluid into the air during the 16 hours the well was out of control. According to a *Pittsburgh Post-Gazette* story, Pennsylvania Environmental Secretary John Hanger ordered EOG and one other company involved in the blow-out to ensure their drilling practices are safe for their workers and the neighboring community. Mr. Hanger also added that if "If EOG

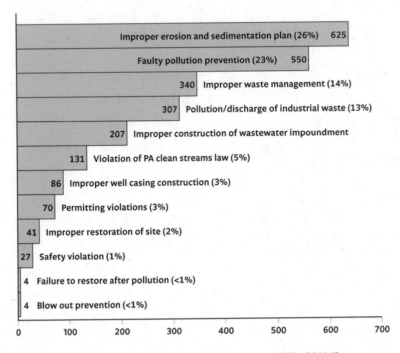

FIGURE 15.04. Environmental Violations by Category 2008–2011.[19]

violates this order, we are headed to terminate their privilege to do business in Pennsylvania."[20]

- In April 2011, at the request of the DEP, Chesapeake Energy shut down all fracturing operations in the state for three weeks after the company lost control of its Atgas 2H well in Bradford County during completion operations. According to the DEP, fluids from the well mixed with rainwater and flowed into a nearby creek. CHK paid a $190,000 fine as a result of the incident.[21]

- In May 2011, Chesapeake Energy was fined $900,000, the largest fine ever handed down by the DEP, for contaminating private water wells in Bradford County due to improper well casing.[22]

- Cabot Oil and Gas was fined $240,000 for its failure to comply with a November 2009 DEP order to plug three wells in Dimock Township, Pennsylvania, that had contaminated the water supplies of 14 residences with gas. The DEP identified poor or improper cement casings were the cause of the contamination.[23]

Does Hydraulic Fracturing Impact Water Wells?

In 2011, Robert Jackson and Avner Vengosh and two other researchers at Duke University published a study linking elevated methane in water wells in Pennsylvania to Marcellus development. The study, which was published online in May 2011 by the Proceedings of the National Academy of Sciences website, found methane levels in water wells located within one kilometer of a gas well that was hydraulically fractured to be up to seventeen times higher than other wells.[24] The Duke researchers also found elevated methane levels in 85 percent of the sixty water wells they tested in northeast Pennsylvania and upstate New York.[25] Figure 15.05 shows the high occurrence of elevated methane levels in water wells (dots) less than 1,000 meters from a hydraulically fractured gas well.

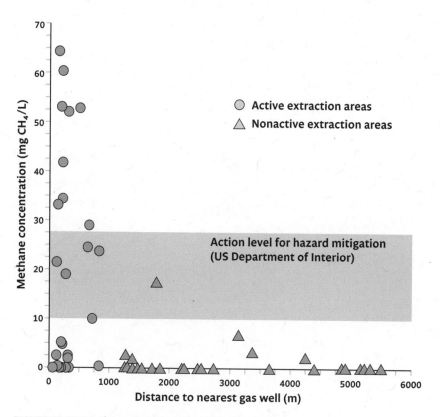

FIGURE 15.05. Methane Concentration Near Gas Wells.[26]

Additionally, the Duke study settled much of the debate over the source of the methane found in contaminated water wells. Prior to the study, it was widely theorized that water wells could only be contaminated with biogenic methane, which is only found in shallow formations, has long been found in water wells throughout the region and is generally not associated with hydrofracking. "Deep gas has a distinctive chemical signature in its isotopes," explained Dr. Robert Jackson, one of the authors of the study. "When we compared the dissolved gas chemistry in well water to methane from local gas wells, the signatures matched."[27]

While little is known about the health effects of methane-contaminated drinking water, the Duke study did note that methane is an "asphyxiant in enclosed spaces and an explosion and fire hazard."[28] The study pointed to leaky well casings as the likeliest cause of methane migration and recommended additional monitoring and sampling of water wells in active drilling areas to better understand the issue. [29]

Remarkably, rather than thanking the researchers at Duke for providing a service to the people of Pennsylvania for a detailed, balanced and well-designed study of methane migration—a very important issue for many in the state—Pennsylvania Environmental Protection Secretary Michael Krancer called the study "biased science by biased researchers."[30] I found the study to be incredibly balanced and fair since it pointed out that while there is evidence of a connection between drilling activity and methane migration, it found no evidence of fracture fluid in water wells. Perhaps Mr. Krancer is the one who is biased?

CHAPTER 16

Arkansas: Limits of the Fayetteville Shale Abundantly Clear

WITH MORE THAN seven years of production data from the Fayetteville Shale of Arkansas and billions in write-downs by two of the play's biggest operators, it is quite clear the play will never meet the expectations of its early promoters.[1] The Fayetteville, which accounts for approximately 75 percent of all gas production in the state of Arkansas, is a layer of dense organic, gas-rich rock that lies approximately 6,500 feet below the surface in the core area of the play.[2] Geologists from Southwestern Energy (NYSE:SWN) first discovered it in 2002 while drilling in a sandstone formation that was unexpectedly productive. This discovery, which became known as the "Weddington Incongruity," is marked on the map below, Figure 16.01, which also displays the positions of both the Barnett and Fayetteville shales along the Ouachita Thrust Belt.

The technical staff at SWN determined much of the excess gas production from the sandstone reservoir on the Weddington Incongruity had migrated up from the Fayetteville Shale directly below. Recognizing the many geological similarities to the Barnett Shale of the Fort Worth Basin, SWN smartly snapped up 343,000 net acres for $11 million in 2003 and acquired an additional 560,000 by the end of 2006.[4] These

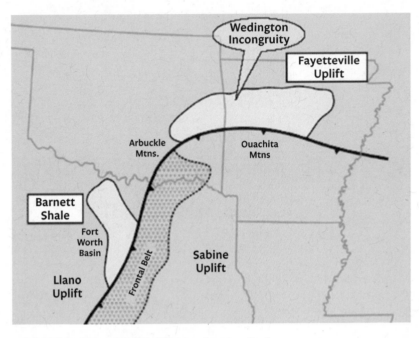

FIGURE 16.01. Map of the Weddington Incongruity.[3]

early acreage acquisitions now form the bulk of the Fayetteville's core area.

Now that the Fayetteville has been producing for over seven years, we can make three important observations based on a meaningful amount of production data. First, it is obvious that though the Fayetteville Shale underlies a large portion of northern Arkansas, a relatively small core area of the play has delivered all of its economic wells. With over four thousand wells drilled in the play by mid-2012, a distinct core area has developed in Conway and Van Buren counties,[5] with a secondary core area in White County. Figure 16.02 shows the drilling through mid-2012 by play leader Southwestern Energy in the Fayetteville.

It should be noted that only wells represented by the red and blue dots—those that had initial production (IP) rates over three million cubic feet per day (mmcf/d)—should be considered economic given the approximately $3 million required to drill and fracture stimulate a Fayetteville well.

The second important conclusion we can draw about the Fayette-

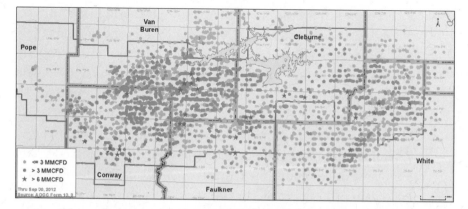

FIGURE 16.02. Southwestern Energy—Fayetteville Shale Focus Area.[6]

ville is that the estimated ultimate recovery (EUR) per well will be far smaller than the rosy predictions put out by its operators. In its August 2010 operations update, Chesapeake Energy announced that the company had moved up its EUR per Fayetteville well from 2.4 billion cubic feet (bcf) to 2.6 bcf due to "continued strong production results."[7] Could Chesapeake, which sold its Fayetteville assets to BHP Billiton for $4.5 billion in February 2011, have made an overly optimistic assumption about the Fayetteville well productivity? I think so. To put into perspective how optimistic CHK's claim of 2.6 bcf was, consider the following: of the 4,258 wells completed in the Fayetteville Shale between January 2005 and the end of July 2012 (by Chesapeake/BHP and other operators), only 1,116 (26 percent) have produced more than 1 bcf and only 86 wells—2 percent—have produced more 2 bcf.[8] Despite the announcement of "continued strong production results" in 2010, the average cumulative production per well amounted to only 719 mmcf for the 887 wells Chesapeake/BHP drilled in the Fayetteville through July 31, 2012.[9] In Q2 2012, less than 18 months after acquiring Chesapeake's Fayetteville assets, BHP wrote off $2.84 billion of its Fayetteville assets, more than a 50-percent reduction from the purchase price.[10]

It is not just Chesapeake that may have overstated recoveries from the Fayetteville. On its Q4 2010 earnings conference call in February 2011, Southwestern Energy upped its EUR for its proven undeveloped Fayetteville drilling locations from 2.2 bcf per well at the end of 2009 to

2.4 bcf per well at the end of 2010.[11] According to data from the Arkansas Oil and Gas Commission (AOGC), between 2005 and the end of July 2012, the company drilled 2,625 wells with an average cumulative production of 804 mmcf.[12] Though some of Southwestern's wells are still early in their productive lives, it is difficult to foresee any scenario under which they will meet their expected 2010 EUR of 2.4 bcf per well.

The third important takeaway from the Fayetteville's seven years of development history is the recognition that today's wells are far different from those drilled during the early days of the Barnett Shale more than a decade ago. The first modern shale gas wells were drilled vertically and were completed with only one fracture or with short horizontal legs and only a handful of fractures. Nowadays, the industry is drilling exclusively horizontal wells with progressively longer lateral legs that can, in some cases, accommodate more than two dozen fracture stimulations. As you can see from Table 16.01, taken from an October 2012 Southwestern Energy presentation, Fayetteville well productivity after sixty days has fallen over the past couple of years despite an increase in average lateral length. Lower productivity from longer lateral wells indicates the company has exhausted its best locations and is drilling into progressively lower-quality rock.

So how much gas will an average Fayetteville Shale well be expected to produce over its lifetime? Based on the production data available from the AOGC, a reasonable estimate would be 1.3 bcf. While there will be plenty of wells that will exceed 1.3 bcf, there have been hundreds of very poor wells that will keep the average well from materially exceeding this figure. Now that we have a reasonable understanding of how much each Fayetteville well will produce, we can make an informed estimate of what the entire play can be expected to produce over its lifetime. Based on a core area of approximately 750,000 acres and 100-acre spacing, it is reasonable to assume the Fayetteville will produce approximately 10 tcf over its lifetime.

With the Fayetteville's rapid maturation and the sharp slowdown in drilling that occurred there between 2008 and 2012, the days of production growth for both this shale and the state of Arkansas are now over. The Fayetteville Shale likely peaked in the second half of 2012 at a rate

TABLE 16.01. Fayetteville Shale — Horizontal Well Performance.[13]

Time frame	Wells placed on production	Average IP rate (Mcf/d)	30th-day average rate (# of wells)	60th-day average rate (# of wells)	Average lateral length
1st Qtr 2007	58	1261	1066 (58)	958 (58)	2104
2nd Qtr 2007	46	1497	1254 (46)	1034 (46)	2512
3rd Qtr 2007	74	1769	1510 (72)	1334 (72)	2622
4th Qtr 2007	77	2027	1690 (77)	1481 (77)	3193
1st Qtr 2008	75	2343	2147 (75)	1943 (74)	3301
2nd Qtr 2008	83	2541	2155 (83)	1886 (83)	3562
3rd Qtr 2008	97	2882	2560 (97)	2349 (97)	3736
4th Qtr 2008[1]	74	3350[1]	2722 (74)	2386 (74)	3850
1st Qtr 2009[1]	120	2992[1]	2537 (120)	2293 (120)	3874
2nd Qtr 2009	111	3611	2833 (111)	2556 (111)	4123
3rd Qtr 2009	93	3604	2624 (93)	2255 (93)	4100
4th Qtr 2009	122	3727	2674 (122)	2360 (120)	4303
1st Qtr 2010[2]	106	3197[2]	2388 (106)	2123 (106)	4348
2nd Qtr 2010	143	3449	2554 (143)	2321 (142)	4532
3rd Qtr 2010	145	3281	2448 (145)	2202 (144)	4503
4th Qtr 2010	159	3472	2678 (159)	2294 (159)	4667
1st Qtr 2011	137	3231	2604 (137)	2238 (137)	4985
2nd Qtr 2011	149	3014	2328 (149)	1991 (149)	4839
3rd Qtr 2011	132	3441	2666 (132)	2372 (132)	4847
4th Qtr 2011	142	3646	2606 (142)	2243 (142)	4703
1st Qtr 2012	146	3319	2421 (146)	2131 (146)	4743
2nd Qtr 2012	131	3500	2454 (121)	2003 (77)	4840

Source: Southwestern Energy.
[1] The significant increase in the average initial production rate for the fourth quarter of 2008 and the subsequent decrease for the first quarter of 2009 primarily reflected the impact of the delay in the Boardwalk Pipeline.
[2] In the first quarter of 2010, the company's results were impacted by the shift of all wells to "green completions" and the mix of wells, as a large percentage of wells were placed on production in the shallower northern and far eastern borders of the company's acreage.

slightly above 2 bcf/d. While I do not expect a rapid decline in production from the shale in the next few years, a modest decline can be expected should Southwestern continue drilling.

Figure 16.03, which charts Arkansas natural gas production history from 1980 through 2010, shows limited but stable production before a significant ramp upwards once the development of the Fayetteville began in 2005.

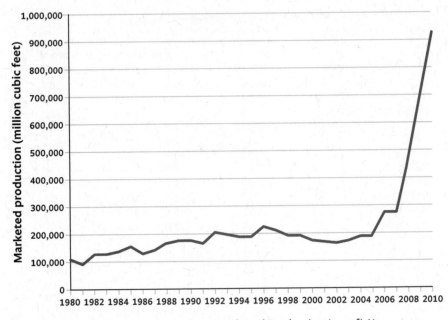

FIGURE 16.03. Arkansas Natural Gas Marketed Production (mmcf).[14]

CHAPTER 17

Canada:
Era of Cheap Exports to
the US Ending Soon

C ANADA HAS BEEN the largest foreign supplier of natural gas to the US by a wide margin for more than three decades. But after more than a decade of falling production and rising demand, the era of Canada exporting cheap natural gas to the US is coming to a close. As I discussed in Part One, despite the discovery of several unconventional resource plays such as the Montney hybrid shale and the very promising Horn River Basin that undoubtedly hold trillions of cubic feet of natural gas, Canada's natural gas production peaked in 2002 and shows no sign of rebounding. In fact, the weak price environment from mid-2008 to 2012 that led to a large reduction in drilling activity has accelerated production declines in recent years. As production wanes, demand is rising to record levels as industrial demand continues to grow along with expanded use from oil sands. In this chapter I will review the changing dynamics of the Canadian natural gas market and put forward compelling evidence that Canada will not be able to provide any relief during the upcoming gas crisis.

Canadian Production

An important—almost never discussed—trend in the Canadian gas market has been the consistent decline in production over the past decade, irrespective of prices. For example, despite a huge ramp-up in drilling between 2002 and 2008 that coincided with prices moving from an average of $3.83CDN per thousand cubic feet (mcf) in 2002 to $7.73CDN in 2008, production dropped 8.5 percent during this period.[1] While much of the increased drilling was directed towards low-productivity Horseshoe Canyon CBM wells in Alberta, falling production during a period when prices nearly doubled shows that conventional Canadian production has entered terminal decline.

Will Canada's emerging plays and the unconventional plays under development be enough to arrest the country's decade-long production slide? Not for a while. With only 83 gas rigs running as of October 12, 2012 and prices still far below the break-even point for virtually every gas play, Canadian production is likely to fall to close to 12 bcf/d by year-end 2012.[2] As you can see from Table 17.01, Canadian production has been falling for a decade.

While Canada's unconventional plays hold trillions of cubic feet of gas—just how much is largely unknown due to limited production history—developing them will take considerable time and cost billions of dollars. Based on a realistic development scenario, it is reasonable to expect that the Montney, a shale/siltstone hybrid play straddling the Alberta/British Columbia border, can grow from its current 1 bcf/d to

TABLE 17.01. Canadian Daily NG Production 2001 to 2012.[3]

Year	Average daily production in bcf/d[1]	% Change from previous year
2001	16.56	NA
2002	16.69	+.01%
2003	16.12	−2.8%
2004	16.14	0%
2005	16.55	+2.5%
2006	16.60	0%
2007	16.17	−2.65%
2008	15.27	−5.6%
2009	14.22	−6.9%
2010	13.96	−1.8%
2011	14.05	+.01%
2012[2]	13.00	−7.5%

Source: Natural Resources Canada.
[1] Billion cubic feet per day.
[2] Author's estimate.

between 2 and 3 bcf/d over the next three to five years. This growth in Montney production assumes gas prices on Canada's AECO Hub—the country's major pricing hub—rise substantially to increase drilling activity. Additionally, it is easy to envision production from the Horn River Basin increasing to 3 bcf/d by 2017, compared to only approximately 350 mmcf/d currently. Figure 17.01 shows Canada's two most important unconventional gas plays, the Montney and the Horn River Basin.

In addition to the play's distance from markets, the big constraint on Horn River development will be the need to build treatment facilities to remove CO_2 from the production stream. Approximately 12 percent of the gas production stream in the Horn River is CO_2.[5] Outside of the Montney and Horn River plays, numerous other unconventional gas plays already under development—such as the Cardium, Viking and

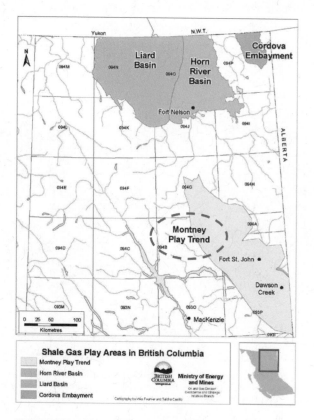

FIGURE 17.01. Map of Montney and the Horn River Basin.[4]

Wilrich formations—are likely to make up an increasing percentage of Canada's gas production over the next decade.

Considering the contributions from Canada's existing and emerging unconventional plays and subtracting out the continuing decline in conventional production, the country's overall production should stabilize somewhere between 10 and 11 bcf/d over the next few years. Unlike the U.K., which is experiencing a production death spiral as its North Sea reservoirs are rapidly depleted, Canada's large endowment of high-quality, undeveloped, unconventional reserves ensures the country will be a major producer of gas for many years.

Canadian Demand

According to the country's National Energy Board (NEB), Canada's consumption of natural gas rose to an all-time record in 2011 of 8.3 bcf/d, up 8 percent from 2010.[6] As you can see from Figure 17.02, similar to US consumption patterns, Canada is witnessing growing industrial de-

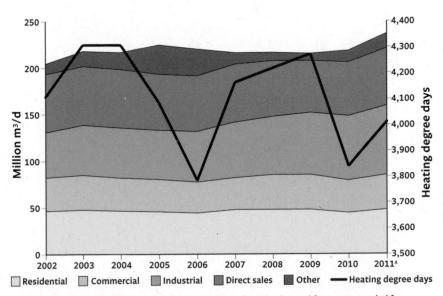

ᵃ Heating degree days (HDD) is an index calculated to reflect the demand for energy needed for heating homes, businesses, etc. HDD is the cumulative number of degrees in a year for which the mean temperature falls below 18.3 degrees C.

FIGURE 17.02. Canadian Natural Gas Demand by Sector.[7]

mand, largely because the country is home to some of the lowest gas prices in the Western world.

One area of industrial demand that should see continued growth over the next several years is the fertilizer sector. For example, in mid-2012 fertilizer maker Yara International announced an expansion of its Canadian urea production facility due to increased demand in the Northern Plains.[8] I expect many more companies to expand or build new plants given the profitability of the Canadian fertilizer business.

However, the biggest driver of industrial demand growth for the next decade will be the oil sands sector. With numerous large-scale oil sands projects already under construction or in the advanced permitting stages, demand from the oil sands will continue to grow from the 1.4 bcf/d that was consumed in 2011 (up 14 percent from 2010 levels).[9] Figure 17.03, taken from the National Energy Board's Canadian Energy Overview 2011, clearly shows the large ramp-up in demand from the oil sands sector over the past decade.

With 787,905 barrels of oil sands capacity under construction in October 2012, a 35-percent increase on existing production capacity

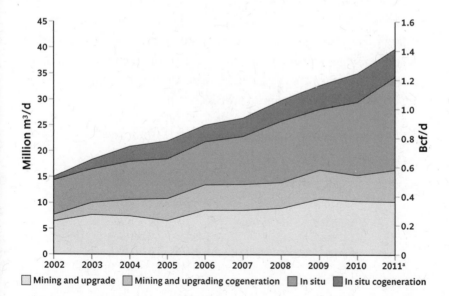

FIGURE 17.03. Average Annual Purchased Natural Gas Requirements for Oil Sands Operations.[10]

of 2,250,400, it is easy to see how demand for natural gas from the oil sands will reach approximately 2 bcf/d by 2015.[11] Additionally, should oil prices remain near $100 per barrel and oil sands take-away issues get resolved in a reasonable amount of time (e.g., with the building of the Keystone XL Pipeline to take oil sands production to US refineries or approval of the Northern Gateway Pipeline to transport production to British Columbia for shipment to Asia), oil sands production could reach more than 4 million barrels a day by 2020. Such a scenario would result in natural gas demand from the oil sands of approximately 2.5 bcf/d.

Adding It All Up

While it's impossible to say exactly when Canada will cease exporting gas to the US, the trend is clear. *Due to Canada's dropping production, increased domestic demand and its ability to export LNG to Asian markets by 2014, the US could see net imports from Canada drop to nearly nothing as early as 2015.* In Table 17.02 I have estimated demand flows for the Canadian gas market for 2012 based on a reasonable extrapolation of current trends.

So how will the US replace declining Canadian imports at a time of falling US production? It won't. US natural gas prices will have to rise to destroy enough demand for the market to reach a new equilibrium. As we know from the 1970s gas crisis, demand destruction through escalating prices can be a very painful process.

TABLE 17.02. Supply and Demand Flows 2012 and Projected 2015

	2012	2015
Supply		
Canadian production	13 bcf/d	11 bcf/d
Demand		
Oil sands	1.5 bcf/d	2 bcf/d
Rest of Canada demand	6.8 bcf/d	8 bcf/d
Net exports to US	4.7 bcf/d	0.3 bcf/d
LNG exports	0 bcf/d	0.7 bcf/d
Total	**13 bcf/d**	**11 bcf/d**

Source: Author's estimates.

CHAPTER 18

LNG Will *Not* Save Us in the Next Crisis

IN RECENT YEARS there has been growing interest in US producers exporting Liquified Natural Gas (LNG), as a way to take advantage of higher overseas natural gas prices. Construction of a new export facility to allow for LNG exports by 2015 at the Sabine Pass LNG Terminal [majority owned by Cheniere Energy (NYSE:LNG)] is already underway, *despite the fact that the US is still a net importer of natural gas.*[1] The 100-year supply myth has clearly been embraced. The belief that the US can displace imports with growing production from the Marcellus and elsewhere appears to have been central to the Federal Energy Regulatory Commission's (FERC) decision to grant licenses to build or convert four facilities to export LNG.[2] The facilities will have the capacity to export 7.6 bcf/d of LNG once complete.[3]

However, irrespective of our export capacity, I do not believe the US will ever export a material amount of LNG. During the next supply crunch, politicians will likely revoke all export licenses, under the guise of *force majeure*, instead of telling suffering constituents that the US needs to honor contractual obligations to overseas customers (overseas customers do not vote). In fact, as natural gas prices reach well into the double digits over the next few years, the US will once again look to the world LNG market for new supplies. The last two times that happened—during the 1970s gas crisis and the period of continuously rising prices between 2002 and 2008—there was little relief, and there won't be next time either. Worldwide demand for natural gas reached a record 114 tcf in 2011, up 31 percent from 2001 levels, and, despite recent export capacity additions from countries such as Qatar (see Chapter Four), the worldwide market for LNG is very tight.[4] In this chapter I will review the supply/demand dynamics of the worldwide LNG market and discuss the difficulties the US will face when it needs to increase imports.

Worldwide LNG Supply

From nearly a standing start a decade ago, the tiny country of Qatar has come to dominate the world LNG trade. As discussed in some detail in Chapter Four, the country has greatly expanded its production, consumption and exports of gas over the past decade,[5] and accounted for a whopping 31 percent of worldwide LNG exports in 2011; the next largest LNG exporter was Malaysia, with only 10 percent of the world total.[6] Table 18.01 lists the world's top exporters of LNG in 2011.

Though it was only fourth in 2011, Australia is set to challenge Qatar as the world's largest exporter of LNG by the end of the decade. According to a recent Reuters article, Australia currently has $170 billion of LNG export projects under construction.[8] The country has plans to increase its export capacity four-fold between early 2012 and the end of 2018, to approximately 3.13 tcf per year, as eight new projects come onstream.[9] Australia's rise to near the top of the LNG export charts

TABLE 18.01. World's 10 Largest LNG Exporters in 2011.[7]

Country	LNG exports in tcf
Qatar	3.623
Malaysia	1.176
Indonesia	1.031
Australia	0.914
Nigeria	0.914
Trinidad and Tobago	0.667
Algeria	0.603
Russia	0.508
Oman	0.384
Brunei	0.331

Source: 2012 BP Statistical Review.

will not be easy or cheap. Slightly more than half of the expansion of the country's LNG export capacity will come from offshore projects off the northwest coast while the balance will source gas from the massive onshore coal seam gas fields in Queensland,[10] so production costs will be relatively high compared to other LNG exporting jurisdictions. The country also faces sky-high material and labor costs, since its LNG boom is occurring simultaneously with a massive increase in its mining capacity. As a result, Australia is the world's highest-cost producer of LNG. Numerous projects currently under construction require a sales price of approximately $10 per thousand cubic feet (mcf) to break even.

While Australia is certain to grow LNG exports and become the most important supplier to the growing Asian market over the next few years, both Trinidad and Tobago and Indonesia—currently the third and fifth largest exporters, respectively—will likely see their exports shrink over the same time period.[11] Let's examine the case for declining exports for Indonesia first.

Despite strong prices for LNG in 2011, Indonesian exports from the country's three export terminals declined 8.1 percent compared to 2010, due to a combination of declining production from its Arun and Bontang export projects and rising domestic demand.[12] Exports for 2012 are expected to fall slightly again despite continued high prices. Though Indonesia was credited with proven reserves of 105 tcf, or over 39 years of production at current rates, as of the end of 2011, a 2006 decision by the government to give domestic customers priority over exports casts doubt over the role Indonesia will play in the growing world LNG trade.[13] Given the country's high population and declining oil production, which has turned it from an exporter of crude to an importer in recent years, I suspect Indonesia will focus on using its large endowment of natural gas to create jobs for its population of over 200 million people rather than maintain its current level of exports.

Trinidad and Tobago has become an important exporter of gas over the past decade as its production has grown from 1.50 bcf/d in 2001 to 3.94 bcf/d in 2011.[14] As you can see from Figure 18.01, both gas production and consumption—mostly due to increased fertilizer production—have risen substantially over the past 12 years.

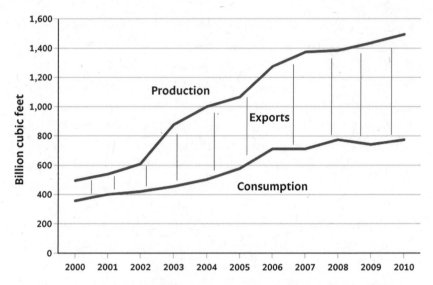

FIGURE 18.01. Natural Gas Production and Consumption Trinidad & Tobago, 2000–2010.[15]

However, tougher times are ahead for the country's natural gas sector. While 2011 production was only down 4.2 percent from the record level of 2010, the country has taken several major reserve write-downs in recent years. According to the *Oil and Gas Journal,* Trinidad and Tobago has seen its proven reserves decline from 25.9 tcf in 2006 to 14.4 tcf in 2011, a drop of 44 percent.[16] Table 18.02 shows the year-over-year decline in the country's proved reserves since 2005.

An August 2011 reserve report by reservoir engineering firm Ryder Scott found that proved reserves had declined further, to 13.46 tcf.[18] With approximately nine years of proved reserves left at today's rate of production, the country is in desperate need of new discoveries. Unfortunately, all 56 wells drilled on the island nation during the first nine months of fiscal 2011 were development wells—wells drilled for

TABLE 18.02. Proved Reserves of Natural Gas in Trinidad & Tobago, 2005–2011.[17]

Year	2005	2006	2007	2008	2009	2010	2011
Trillion cubic feet	25.887	25.88	18.77	18.77	18.77	15.4	14.416
Percentage change YoY	0%	0%	−38%	0%	0%	−22%	−7%

Source: EIA.

the purpose of converted proved undeveloped reserves (PUDs) into proved developed producing reserves (PDPs).[19] However, all hope is not lost. With the awarding of four shallow-water blocks in November 2010 and two deepwater blocks in July 2011, new discoveries may be on the horizon.[20]

While there is certainly significant LNG export capacity coming on-line before the end of the decade from countries such as Australia and Algeria, expansions in these countries and elsewhere will be partially offset by declines in exports from Indonesia, Trinidad and Tobago and Middle Eastern countries such as Oman and the United Arab Emirates (UAE). I will discuss the outlook for LNG exports from Oman and the UAE below.

Worldwide LNG Demand

I see three big trends influencing the demand side of the world LNG market of the next five years. One of the most significant movements currently underway is the shift in LNG demand from Europe to Asia. The combination of contracting economies, increased renewable capacity and increased utilization of coal-fired generation has dampened LNG demand in Europe.[21] Europe's accelerating economic slump has pushed down the price of coal and carbon emissions (Europe has a carbon permit program), making gas-fired generation expensive by comparison. Utility companies from Germany, Britain, Italy and the Iberian peninsula have reduced their reliance on natural gas as a feedstock due to the price advantages of coal.[22] Meanwhile Asia, home to the first and second biggest importers of LNG, Japan and South Korea respectively, continues to expand its imports. For example, for the first six months of 2012, LNG cargoes to Continental Europe were down 33 percent and down a whopping 50 percent to the UK while imports into Asia were up 16 percent.[23] While demand from traditional Asian importers remains strong, China is quickly developing a gluttonous appetite for LNG; the sixth largest importer of LNG in 2011, the country is building more than six new LNG terminals that will double its import capacity by 2014.[24] With pollution out of control due to its enormous fleet of coal-fired power plants, there is little doubt that China will challenge Japan for the

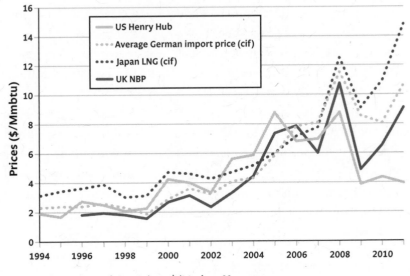

FIGURE 18.02. Natural Gas Prices $/Mmbtu.[26]

title of the world's biggest LNG importer in less than five years time as it shifts towards cleaner-burning natural gas.

Despite its moribund economy, Japan continues to set new records for LNG consumption. The country's imports have shot up since its nuclear generation industry was shut down following the March 2011 Fukushima–Daiichi disaster. Japan imported a record 10.35 bcf/d (3.78 tcf) of natural gas via LNG in 2011, up from 9.21 bcf/d (3.36 tcf) in 2010, an increase of 11.1 percent.[25] Figure 18.02 shows the country paid a record amount for natural gas imports in 2011.

There was no relief to Japan's record high prices in 2012. Due to competition for spot cargoes and high Brent crude prices (spot LNG prices are priced off of oil prices), the country paid approximately $18 per MMBtu (one Mcf = 1.023 MMBtu) in July 2012 before prices backed off in early August.[27] Japan's elevated level of LNG imports is likely to continue for some time since the country has been very slow to return nuclear reactors to service—the first one came back online in July 2012—due to public safety concerns.[28] Figure 18.03 is a chart that includes Japanese LNG import prices for the first eight months of 2012.

The second major swing in demand for LNG over the next few years is increasing demand from South America. Argentina's price for landed

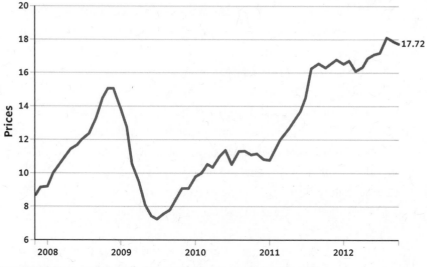

FIGURE 18.03. Japan Liquefied Natural Gas Import Price Chart.[29]

LNG was the highest in the world in August 2012 and Brazil's was the fifth highest. Figure 18.04 shows how Southern American LNG prices were among the highest in the world in August 2012.

While the high South American prices can be partially explained by winter weather in the Southern Hemisphere, there is little doubt that a major upswing in demand is occurring in the region. According to the

FIGURE 18.04. World LNG Estimated August 2012 Landed Prices.[30]

2012 BP Statistical Review, Central and South American countries increased their imports of LNG from 325 bcf in 2010 to 385 bcf in 2011, an increase of nearly 23 percent.[31] Though South America is not yet a large LNG importer by world standards, this may change should recent demand trends continue. For example, Argentina is set to import a record amount of LNG in 2012, despite a large increase in imports of Bolivian gas. The country, which has seen gas production decline 16 percent from its peak level in 2004, is set to pay an average of $10 per MMBtu for piped in Bolivian gas versus $16 per MMBtu for LNG.[32] Unless Argentina can halt the decline in its natural gas production it, along with Brazil and Chile, will continue to increase LNG imports for the next several years.

The third and potentially biggest driver for LNG demand over the next decade will be increased demand from the US. While US LNG imports in 2011 totaled only 348 bcf—the lowest since 2002, when 228 bcf was imported—we are likely to see a dramatic rebound in US reliance on LNG over the next several years.[33] Given the huge drop in shale gas-directed drilling, falling production in the Gulf of Mexico and major states like Louisiana, New Mexico, Texas and Wyoming and the drop in Canadian imports, it is easy to foresee a scenario where the US would have to tap world LNG markets for five percent of supply by 2015. Five percent of 2011 total US gas consumption amounts to approximately 3.3 bcf/d or 1,215 bcf per annum, or 2.4 bcf more LNG than the US imported on a daily basis in 2011.[34] Finding an extra 1.2 tcf per year in a global market of only 11.7 tcf in 2011 will be an extremely difficult task.[35]

Will LNG Save the US in the Next Crisis?

Despite LNG import capacity of 19 bcf/d, up from approximately 4 bcf/d at the turn of the new millennium, the US has very little chance of materially increasing its LNG imports during the upcoming natural gas crisis.[36] The problem is twofold. First, US terminal operators source most of their cargoes on the spot market rather than signing long-term contracts that guarantee supply. While this practice has saved many operators from buying expensive cargoes they would have been forced to sell at a loss during the period of weak prices between 2009 and 2012, it did not increase LNG imports in 2008 at a time of record prices and it will not work during the next gas crunch. In fact, between 2007 and 2008, LNG imports *declined* from 2.11 bcf/d in 2007 to .96 bcf/d in 2008, a 55-percent drop, despite an average wellhead price that rose 28 percent, from $6.25 to $7.97 per mcf.[37]

The second impediment to increased LNG importation is the US's historical reliance on Trinidad and Tobago for the majority of its imports. According to the US Energy Information Agency (EIA), between 2006 and 2011 the US imported an average of 1.34 bcf/d of LNG, 57 percent or 0.76 bcf/d from Trinidad and Tobago.[38] Such heavy reliance on a country with a proven reserve life of only nine years and few new discoveries is a very dangerous game. Without new discoveries in the very near future, Trinidad and Tobago will certainly be producing and exporting substantially less gas by 2020.

So where will the incremental LNG come from during the next natural gas crunch? Probably Middle Eastern countries such as Algeria or Yemen or Australia, but the US will face stiff competition for any available cargoes. As explained in Chapter Four, both Kuwait and the Emirate of Dubai began importing LNG recently (in 2009 and 2010 respectively), to run their desalination and industrial plants as well as gas-fired power plants.[39] Another member of the United Arab Emirates (UAE), the Emirate of Fujairah, is developing a floating LNG terminal on the Gulf of Oman so it can import LNG without requiring ships to enter the Strait of Hormuz. The floating facility is expected to be operational by 2015.[40] The UAE exported LNG for decades but saw its exports decline to nearly nothing by the end of the first decade of the twenty-first century.[41] Neighboring Gulf countries Bahrain and Oman are also expected to announce plans to install initial LNG import terminals.[42]

The LNG business in Oman is very reflective of the state of the natural gas business in the Middle East. Oman, which has three export terminals and was the ninth-largest exporter of LNG in 2011, has been reducing exports since 2007.[43] To meet rising domestic demand and turn around its declining LNG exports, the country recently signed up BP to spend $24 billion to develop a tight sands field through the use of horizontal wells and hydraulic fracturing.[44] Given BP's near total wipe-out in their US shale gas efforts, I am very skeptical the company will have more success in developing a potentially large but technically challenging field that was discovered in the 1990s and remained undeveloped for nearly twenty years.

Will the US be able to tap into Australia's growing LNG exports?

Given that Australia has already committed most of its current capacity and incremental production to long-term contracts and the more than 35 days it takes for Australian gas to reach American import terminals, any LNG imports from the Land Down Under will be rather expensive.

In conclusion, the world LNG market is likely to remain very tight. Much of the current global supply is already accounted for under long-term supply contracts and securing cargoes on the spot market will likely prove difficult. Despite its 19 bcf/d of import capacity, the United States is unlikely to rely on LNG for even five percent of its supply any time soon.

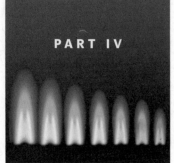

PART IV

WHAT
SHOULD BE DONE

IN THE FOURTH and final part of this book, I will briefly examine the history of the US Energy Information Administration (EIA) and its contribution to the 100-Year Supply Myth, explore practical solutions to the upcoming natural gas crisis and provide proof that the period of low gas prices between 2009 and 2012 was an aberration. In Chapter 19 I look at how the EIA's failure to modernize its monthly US natural gas production report (the EIA 914 report) has led to a significant divergence between the production statistics reported by the Administration and those of state oil and gas agencies. Additionally, I examine how the EIA's reliance on outside consultants and poor research methods have led to the agency grossly overstating America's shale gas resources. Many in the media, industry and policymaking circles incorrectly view the EIA's unrealistic estimates of shale gas resources as validation of the shale gas supply myth.

Chapter 20 is devoted to mitigating the coming natural gas deliverability crisis. Surprisingly, there are numerous ways we can reduce the impact lower natural gas supplies will have on our daily lives and the economy. Though alternative energy companies have struggled tremendously in recent years, the coming era of substantially higher natural gas prices will reverse the fortunes of the survivors and encourage many new start-ups. Using proven, off-the-shelf technologies and a few emerging ones, we can significantly reduce our natural gas consumption. I also introduce the forward-thinking approach the Netherlands has applied to the development of that country's substantial natural gas resources. The Dutch have smartly set production limits that have ensured that natural gas is produced for the maximum benefit of all stakeholders. Sometimes more is not better.

The final chapter is an alternative look at the future. I discuss how many of the factors that pushed gas prices to decade lows in 2012 are now reversing. Unbeknownst to many, horizontal drilling and hydraulic fracturing are not new technologies. I examine the application of these two technologies to the Austin Chalk play of Texas and Louisiana—one of the first in the country to be developed with horizontal wells—in the late 1980s and the drastic fall-off in production from the play's biggest field once drilling stopped. While the Austin Chalk is very different from modern shale gas plays, it is an excellent example of the limits of new technology. I leave you with what the world might look like should the US experience a 1970s-style natural gas crisis.

CHAPTER 19

The US Energy Information Administration's Mission Failure

THE US ENERGY INFORMATION ADMINISTRATION (EIA), the statistical and analytical division within the US Department of Energy, has continuously overestimated the potential of shale gas. In this chapter I will examine the EIA's mission, its history of failure to accurately estimate US monthly natural gas production and how its inability to perform an independent analysis of facts about potential shale gas has led to both bad policy and bad investment decisions on the part of industry. Lastly, I will suggest how to reform the EIA and its methodologies for collecting data and making projections.

The EIA was created as part of the Department of Energy Organization Act of 1977 and is the federal government's primary authority on energy statistics and analysis.[1] Here is how the Administration describes its mission, according to its website:

> EIA collects, analyzes, and disseminates independent and impartial energy information to promote sound policymaking, efficient markets, and public understanding of energy and its interaction with the economy and the environment. EIA is the Nation's premier source of energy information and, by law, its data, analyses, and forecasts are independent of approval by any other officer or employee of the US Government.[2]

To carry out the above mission, Congress allotted the Administration a fiscal 2011 budget of $95 million.[3] The EIA employs approximately 370 workers.[4] Among its many duties, it performs two very important functions related to understanding US natural gas deliverability: preparing and publicly releasing its monthly "914 production report"—a snapshot of monthly US natural gas production—and making long-term projections about natural gas supplies.

Since the EIA began issuing the 914 report in 2005, it has had a long history of overestimating US natural gas production. This chronic overestimation was very well documented in an April 2010 *Wall Street Journal* article by Carolyn Cui titled "US Natural Gas Data Overstated." Below is an excerpt from the article:

The Energy Department is preparing to make sweeping revisions to its US natural-gas production data after finding it has been overstating output, raising new questions about the government's collection of energy information.

The monthly gas-production data, known as the 914 report, is used by the industry and analysts as a guide for everything from making capital investments to predicting future natural-gas prices and stock recommendations.

But the Energy Information Administration, the statistical unit of the Energy Department, has uncovered a fundamental problem in the way it collects the data from producers across the country—it surveys only large producers and extrapolates its findings across the industry. That means it doesn't reflect swings in production from hundreds of smaller producers.

The EIA plans to change its methodology this month, resulting in "significant" downward revisions in some areas, according to Gary Long, the acting director of the 914 form, who led the review…

"*The model we have now overestimates production,*" Mr. Long said in an interview. [Emphasis added.] He said the review was prompted by the EIA noticing aberrations in some states. "We saw some numbers we didn't like in Texas; we thought they were a little too high," Mr. Long said....

In December, the Agency reported total new gas supply at 87.8 billion cubic feet a day and total demand of 80 billion, leaving 7.8 billion cubic feet unaccounted for—a margin of error of 10%.

"It's getting ridiculously large," said Ben Dell, an analyst with Sanford C. Bernstein. "*When you have a 10% gap, that's somewhat making a mockery of the data.*" [Emphasis added.]

Mr. Dell in January wrote a report raising questions about the mismatch. In that report, he focused on October numbers that showed a 12% margin of error.

"We think that most would agree that a 12% margin of error makes a data set tough to rely on, to say the least," Mr. Dell wrote in that report. Rather than gas supply being flat or slightly down as the data suggests, Mr. Dell wrote, he believes production is actually falling.[5]

In April 2010, the EIA attempted to improve the accuracy of its 914 report after it found it had included data that was more than two years old, leading to gross production estimation errors (the Administration now uses data between six months and two years old to come up with its production estimates).[6] Despite these efforts, the report remains fundamentally flawed for several important reasons.[7] First, to compile the report, the EIA relies on a private data source along with a voluntary sampling of only about 250 large natural gas producers (companies that produce over 20 mmcf/d), out of the more than 6,000 producers nationwide.[8] It also uses state production data, even though several states, such as Oklahoma, lag many months behind in collecting accurate production data, making the state data in the EIA model very suspect.

A second problem with the EIA's 914 report is its inability to accommodate rapid changes in natural gas production by reporting companies. For example, the EIA's production estimation model often provides grossly inaccurate results when drilling and production ramp up or drop off over short periods of time. Should a large company sell off a major natural gas field, merge with another company or go out of business, the EIA's model is not able to take into account these changes and may produce inaccurate supply estimates. The EIA even admits that its model is ill prepared to handle rapid changes in production in the industry. Here is how the EIA describes these two flaws in its estimation methodology:

> This method is a significant improvement over previous methods in making use of information that is as current as possible. Even so, the historical data still have some lag and so some potential for error remains. For example, the rapid development of the Haynesville shale in Louisiana caused a change in the State production trend that, in turn, may cause the simple ratio method to over estimate. Also, a sample affected by improperly handled mergers or property sales, as described above, can adversely affect the production estimates. EIA continues to look for better ways to handle the mergers and property sales.[9]

Based on the initial production estimates from the Texas Railroad Commission for 2012, it appears the EIA continues to significantly overestimate production from Texas.[10]

Since the EIA began issuing the 914 monthly production report in 2005, policymakers and investors have relied on it for making decisions but have failed to get an accurate reading of America's natural gas production. The EIA's attempt to reform the report has failed. It is time for the EIA to enter the twenty-first century of real-time data collection. The Administration should work with the Department of Transportation's Office of Pipeline Safety to install off-the-shelf wireless monitors at each of America's pipeline tap points that would send real-time production data to a central database (pipeline tap points are facilities that connect a producer's gathering system to the interstate pipeline system). Due to the automated nature of the wireless data collection system, the EIA would be able to publish *actual* monthly production numbers shortly after each month-end. While there will undoubtedly be costs associated with establishing the wireless production reporting network, the cost savings over time to the US Department of Energy would likely be meaningful. Since we now live in an era where we are able to get monthly sales numbers for auto companies, teen retailers and many other industries, the time has come for accurate and timely production numbers and an end to the age of model-produced guesswork. Natural gas is too important to the US economy to not have accurate monthly production numbers.

The EIA's Projection Failures

I will be the first to admit that making projections about the future of America's natural gas deliverability is difficult. Many factors, such as economic activity, advances in technology, capital availability and even the weather, influence the country's ability to produce natural gas. However, the EIA, which makes both short-term and long-term projections on America's natural gas deliverability and other topics—such as the importance of shale gas—has provided numerous projections throughout its history that were based on little if any independent research and few historical facts.

For example, one of the most important data items supporting the myth that natural gas will remain plentiful and cheap for years to come is the EIA's Annual Energy Outlook (AEO). In the 2011 Annual Energy Outlook Early Release, the EIA projected the US contained technically recoverable unproven shale gas resources of 827 tcf.[11] This estimate was up from its 2010 reference case estimate of 347 tcf.[12] Figure 19.01—taken from the 2011 Annual Energy Outlook Early Release—clearly shows the EIA's vision that increasing shale gas production will not only replace a decline in imports but will also support modest production growth until 2035.

Unfortunately, the EIA fails to provide data in its Annual Energy Outlook to support its conclusions on future deliverability or the size of America's natural gas resource base. For example, readers were left guessing at how the Administration arrived at its estimate of 827 tcf of technically recoverable unproven shale gas resources in 2011. However, when the EIA released its "Review of Emerging Resources: US Shale

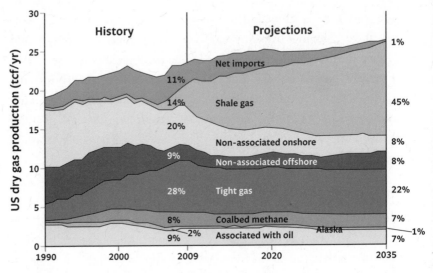

FIGURE 19.01. US Dry Gas Consumption by Production Type.[13]

Gas and Shale Oil Plays" in July 2011, details on the agency's logic became clearer.[14] As the title suggests, this document provides an overview of all oil and gas shale plays in the US and a partial explanation for the EIA's arrival at such a significant natural gas resource base. I find it to be chronically flawed for several reasons.

First, the "Review of Emerging Resources: US Shale Gas and Shale Oil Plays" was not prepared by the Energy Information Administration. The EIA commissioned independent consulting firm Intek, Inc. to prepare the report. So who is Intek and why are they performing a task that is fundamental to the mission of the EIA? Intek is a private consulting firm established in 1998 that has done extensive work for the Department of Energy and other governmental agencies.[15] While some may not see a problem in using an outside consulting firm in report preparation, I do. The "Review of Emerging Resources: US Shale Gas and Shale Oil Plays" is central to the mission of the EIA and should not have been outsourced. Unlike a private consulting firm, the EIA can use its weight as an agency of the federal government to extract information from private sources that might be unwilling to discuss confidential production information with a private firm for competitive reasons. Another reason I disagree with the outsourcing of work that is critical to the EIA's mission is the potential for conflicts of interest with the consulting firm's other clients. This issue was highlighted in a June 26, 2011 *New York Times* article by Ian Urbina titled "Behind Veneer, Doubt about Future of Natural Gas."[16] Urbina attached to the article numerous emails from EIA employees that questioned many of the agency's official conclusions on the future of shale gas and provided the following commentary about the use of outside consultants in the preparation of reports:

> "E.I.A.'s heavy reliance on industry for their analysis fundamentally undermines the agency's mission to provide independent expertise," said Danielle Brian, the executive director of the Project on Government Oversight, a group that investigates federal agencies and Congress.
>
> "The Chemical Safety Board and the National Transportation Safety Board both show that government agencies can conduct complex, niche analysis without being captured or heavily relying upon industry expertise," Ms. Brian added, referring to two independent federal agencies that conduct investigations of accidents.[17]

Second, while it is unclear whether Intek had any conflicts of interest that compromised its objectivity in the preparation of the "Review of Emerging Resources: US Shale Gas and Shale Oil Plays," it certainly relied heavily on industry sources to produce a report that was largely void of independent data collection or analysis. (I contacted Intek numerous times to discuss their report and any conflicts of interest but was unable to speak to one of the firm's principals as this book goes to print.) For example, instead of collecting data on wells to determine appropriate decline curves and estimated ultimate recoveries (EURs) for the various shale plays in its study, Intek simply cut and pasted material from the investor presentations of several shale gas producers. Figure 19.02 appeared on page 8 of its report.

Not only did Intek use a "pro forma" Marcellus well type curve from an investor presentation from Chesapeake Energy *but it used one from the company's 2008 Investor and Analyst Meeting.* Given the limited

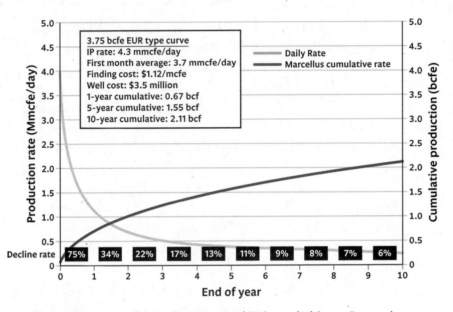

FIGURE 19.02. Marcellus Decline Curve and Estimated Ultimate Recoveries Example.[18]

production history available from the Marcellus in 2008, using a pro forma well type curve which projects Marcellus wells to have an estimated ultimate recovery (EUR) of 3.75 bcf in a report published in July 2011 shows just how little care or effort Intek and the EIA put into their report on the potential of shale gas plays.

Third, Intek/EIA's "Review of Emerging Resources: US Shale Gas and Shale Oil Plays" grossly overstates technically recoverable shale gas resources. Central to its estimate was a table on page 5 of the report (reproduced here as Table 19.01) that listed all the identified shale plays in the US and the amount of gas contained in each play.

It is very clear that Intek and the EIA gave little credence to historical drilling data when preparing and including the above mentioned table in the report. It is difficult to understand how the EIA could publish a report in July 2011 based on "estimates of technically recoverable shale gas and shale oil resources remaining in discovered shale plays as of January 1, 2009" (see date on Table 19.01). Publishing shale gas projections that are two and half years old was a disservice to readers of the report since it ignores a great deal of production history and significant

TABLE 19.01. Intek Estimates of Underdeveloped Technically Recoverable Shale Gas and Shale Oil 1/1/2009.[19]

Onshore lower-48 oil and gas supply submodule region	Shale play	Shale gas resources (tcf)	Shale oil resources (bbls)
Northeast	Marcellus	410	—
	Antrim	20	—
	Devonian Low Thermal Maturity	14	—
	Great Siltstone	8	—
	Big Sandy	7	—
	Cincinnati Arch	1	—
Subtotal		472	—
Percent of total		63%	—
Gulf Coast	Haynesville	75	—
	Eagle Ford	21	3
	Floyd-Neal & Conasauga	4	—
Subtotal		100	3
Percent of total		13%	14%
Mid-Continent	Fayetteville	32	—
	Woodford	22	—
	Cana Woodford	6	—
Subtotal		60	—
Percent of total		8%	—
Southwest	Barnett	43	—
	Barnett-woodford	32	—
	Avalon & Bone Springs	—	2
Subtotal		76	2
Percent of total		10%	7%
Rocky Mountain	Mancos	21	—
	Lewis	12	—
	Williston-Shallow Niobraian	7	—
	Bakken	—	4
Subtotal		43	4
Percent of total		6%	15%
West Coast	Monterey/Santos	—	15
Subtotal		—	15
Percent of total		—	64%
Total onshore Lower-48 States		750	24

Note: From previous EIA estimates and thus not assessed in the INTEK shale report. Subtotals and total may not equal sum of components due to independent rounding.

knowledge gained within the industry in the meantime. As I discussed in Part One, as of mid-2011, there was substantial production data, publicly available, to make informed projections about technically recoverable natural gas resources from several shale plays. Let's examine the differences between my estimate of reasonable future recoveries and the projections of the EIA/Intek of technically recoverable resources from three shale plays with significant production histories (Note: I strongly disagree with the use of the term "technically recoverable resources" in any official resource estimate since it gives credence to a largely useless estimation of future gas production):

1. The Barnett Shale: As I discussed in Chapter 5, by mid-2012, the Barnett Shale had produced approximately 10 tcf of natural gas from nearly 16,000 wells. Given that operators had already established the outer limits of the play, I projected the Barnett may recover another 25 tcf over its lifetime. Intek estimated remaining technically recoverable resources for the play to be 43 tcf (see Table 19.01).[20]

2. The Fayetteville Shale: According to the production history available on the website of the Arkansas Oil and Gas Commission, the play produced 3 tcf of gas from slightly more than 4,000 wells between 2005, when production commenced, and mid-2012. Based on a core area of only 750,000 acres, a realistic estimate of lifetime production from the Fayetteville would be 10 tcf. As you can see in Table 19.01, however, Intek estimates the play has remaining technically recoverable resources (TRR) of 32 tcf—more than three times my estimate.[21] The only justification Intek provides in the report for such a large projection is its estimation that the play is prospective over 9,000 square miles.[22] However, based on nearly seven years of publicly available drilling results, the Fayetteville Shale appears to be prospective over only 1,200 square miles (750,000 acres), a difference of 86 percent.

3. The Antrim Shale: According to Michigan Public Service Commission, the Antrim Shale produced 3 tcf from approximately 10,000 wells between 1989 and the end of 2010.[23] Despite the large rise in natural gas prices between 2000 and 2008, and the many advances in shale gas technology since the play went into commercial pro-

duction in 1989, *production from the Antrim Shale has dropped every year since 1998.*[24] Despite 13 years of decline, and production of only 3 tcf to date, Intek estimated the Antrim Shale contains another 20 tcf of resource potential (Table 19.01). Once again, Intek and the EIA are wildly optimistic in their estimation of the potential of the Antrim based on 22 years of production history. I believe a more reasonable, though very generous, estimate for the potential of the Antrim would be 2 tcf—one-tenth of Intek's number.

In addition to grossly overstating the potential of the country's three shale plays with the most production history, Intek/the EIA also provide optimistic estimates of the potential of the Marcellus and Haynesville shales. According to Intek/the EIA, the Haynesville is expected to recover another 75 tcf of natural gas (Table 19.01). Similar to its analysis of the Fayetteville, Intek/the EIA suggest the Haynesville extends over 9,000 square miles.[25] (The report is unclear about the criteria used to establish the prospective areas for both the Haynesville and the Fayetteville shales.) By mid-2011, the time of publication, approximately 1,000 wells had been drilled into the Haynesville and there was sufficient evidence to determine that a more reasonable estimate of the total play size is approximately 1,562 square miles (1 million acres), with a core area—where virtually all the successful wells to date have been drilled— of between 156 and 234 square miles (100,000 and 150,000 acres). The Haynesville achieved a peak production rate of approximately 6 bcf/d by early 2012. However, this production rate is unsustainable due both to the high decline rate of the wells and the previously discussed 2011 LSU study, which concluded that the average well would produce only 3 bcf over its lifetime. Therefore a generous estimate of lifetime production from the Haynesville would be somewhere between 15 and 20 tcf, not the 75 tcf estimated by Intek/the EIA.

Far and away the most egregious misjudgment of potential deliverability in the Intek/EIA report was its estimate that the Marcellus Shale contained 410 tcf of technically recoverable natural gas resources.[26] The report estimated that the play covers approximately 95,000 square miles and is divided between an active area (defined as acreage that has been

TABLE 19.02. Marcellus Average EUR and Area.[29]

	Active	Undeveloped
Area (sq. miles)	10,622	84,271
EUR (Bcfe/well)	3.5	1.15
Well spacing (wells/sq. mile)	8	8
TRR (Tcf)	177.90	232.44

leased for development) of 10,622 square miles and an undeveloped area of 84,271 square miles (classified as prospective but not currently leased).[27] Intek/the EIA estimates the active area has a technically recoverable resource (TRR) of 177.9 tcf and the undeveloped area contains a whopping 232.44 tcf.[28] Table 19.02, reproduced from the report, further dissects how these numbers were derived.

I find several problems with Table 19.02. First, its numbers do not make sense. Should 8 wells per square mile eventually get drilled into the active area (nearly 85,000 wells)—a highly unlikely scenario—and each well recovers the above-mentioned 3.5 billion cubic feet equivalent (bcfe), TRR from the Marcellus active area would be approximately 297 tcf. Intek provided no explanation on the methodology it used to arrive at a TRR of 177.9 tcf for the active area. Even more troubling was its estimate of 232.44 tcf of TRR in the undeveloped category of the Marcellus. To estimate that the unleased area of the play could eventually support the drilling of 674,168 wells (8 wells per square mile) *and* recover 232.44 tcf is absurd. Once again, the TRR for the undeveloped portion of the Marcellus does not correspond to the estimated number of drilling locations and EUR per well in Table 19.02.

Since the report was published in July 2011, the EIA—in its Annual Energy Outlook (AEO) 2012 Early Release Overview, released in January 2012—has significantly revised downward its estimate of the potential recoveries from the Marcellus, *to only 141 tcf, a decline of 65 percent*.[30] Also included in the 2012 AEO Early Release was a major downward revision of total technically recoverable unproved shale gas resources, from 827 tcf in AEO 2011 to only 482 tcf, *a reduction of 42 percent*.[31] Though the 2012 estimate suggests the US now has unproved shale gas resources of approximately twenty years of US consumption at current rates, the EIA is still likely overstating future shale gas recoveries by a factor of more than three.

Another area where Intek and the EIA over-inflate the importance of shale gas in their report is in its representation of the potential of shale gas plays not currently under commercial development. They estimate that the Barnett–Woodford Shale play of the Permian Basin in west Texas has 32 tcf of technically recoverable resources (TRR) and that the New Albany Shale of the Illinois Basin has 11 tcf of TRR, though neither has established commercial production.[32] If commercial gas cannot be recovered using today's technology, how can Intek/the EIA have confidence in assigning technically recoverable resources to the Barnett–Woodford and the New Albany shales? Even assuming that today's technology would allow for commercial production from these two potentially large plays, nowhere does Intek/the EIA provide the natural gas price that would be needed to turn them into producing fields.

The EIA's Mea Culpa

The first admission from the EIA that its "Review of Emerging Resources: US Shale Gas and Shale Oil Plays" overstated the potential of America's shale gas resources came in August 2011, only one month after the report was published. It was at this time that the US Geological Survey (USGS) released its "New Assessment of Gas Resources in the Marcellus Shale, Appalachian Basin." The report from the USGS concluded the following:

> "*The Marcellus Shale contains about 84 trillion cubic feet of undiscovered, technically recoverable natural gas* and 3.4 billion barrels of undiscovered, technically recoverable natural gas liquids...." (Emphasis added.)[33]
>
> Source: US Geological Survey

Remarkably, the USGS estimate of technically recoverable shale gas resources in the Marcellus is 80 percent lower than the one published by Intek and the EIA. The response from the EIA to the new estimate was somewhat shocking. According to an article published by Bloomberg News, the Administration did not even attempt to defend its estimate of 410 tcf of TRR in the Marcellus. Below is an excerpt from the article:

The formation, which stretches from New York to Tennessee, contains about 84 trillion cubic feet of gas, the US Geological Survey said today in its first update in nine years. That supersedes an Energy Department projection of 410 trillion cubic feet, said Philip Budzik, an operations research analyst with the Energy Information Administration.

"*We consider the USGS to be the experts in this matter,*" Budzik said in an interview. "*They're geologists, we're not.* We're going to be taking this number and using it in our model." [Emphasis added.][34]

Two items jumped out at me when I read Budzik's remarks. First, he was somewhat misleading in suggesting that the EIA does not employ geologists. I have verified through several sources that the EIA does in fact employ geologists. Second, I cannot understand why the EIA, which bills itself as the "Nation's premier source of energy information," is not an expert on the subject of shale gas given the resource's growing importance to America's energy mix.

Can the EIA be Fixed?

So how does the EIA fix itself? First, the Administration should adopt a more practical approach to both the collection and the release of natural gas production data. As discussed above, real-time wireless production data collection monitors at each of America's pipeline tap points and the publication of prior month production data shortly after month-end would provide meaningful insight into the country's deliverability potential. Additionally, the Administration should hire the required technical expertise to allow for the internal preparation of major reports and eliminate the use of outside consultants. Tasked with the important work of providing lawmakers with accurate and timely energy data and analysis, and with an annual budget of nearly $100 million, the Administration certainly has the resources to carry out its mission.

However, the real question is whether there exists the political will in Washington to reform the Administration. Maybe a better solution would be to disband the EIA altogether, assign the Federal Energy

Regulatory Commission (FERC)—the federal regulator of our nation's pipeline system—responsibility for the timely collection and publication of accurate natural gas production data, and then establish a small but highly technical team of petroleum engineers, geologists and statisticians inside the Congressional Research Service to carry out many of the other current duties of the EIA and provide guidance to Congress on important energy issues.

CHAPTER 20

Solutions
to the Coming
Deliverability Crisis

W HILE THE COMING natural gas deliverability crisis will take many market participants, policy makers and consumers by surprise, and is likely to inflict substantial damage to the US economy—potentially even more than the gas crisis of the 1970s—there are numerous ways to mitigate the fallout. Many will be unpopular, all will take time to implement and some will be quite costly. As in the 1970s gas crisis, which was extended and exacerbated by poor leadership in Washington, the government will be of little assistance next time around. In fact, the federal government's efforts to implement short-term fixes to a structural problem are likely to make things worse. In addition to the conservation efforts of the manufacturing sector that are certain to occur with the onset of the deliverability crisis, I have identified six initiatives that will mitigate it. They are as follows:

1. Solar, Wind and Geothermal Going Mainstream

For decades, we have heard that alternative energy sources of heating, cooling and generating electricity have been too costly to implement and would never survive without significant government subsidies. The spate of solar and green-tech bankruptcies of recent years has only reinforced this view. However, this is about to change. Forces pushing from opposite directions will make alternative energy an outstanding choice—even without government subsidies—to power our economy at times of skyrocketing natural gas prices. As I have discussed previously, electricity prices throughout much of the country are based off the price of natural gas. Therefore, the coming spike in natural gas prices will reset electricity rates significantly higher. Because of recent efficiency gains and manufacturing cost reductions, many forms of alternative energy are becoming competitive with traditional sources. For example, a 93-percent collapse in the price of polysilicon—the main ingredient used in solar panels—between 2008 and 2011 has accelerated the move towards grid parity for the solar industry.[1] To take advantage of these prices, more than a dozen solar panel leasing companies are now offering to install panels on homeowners' roofs with no money down and a monthly payment that is usually less than the homeowner's previous utility bill.[2]

While there is little doubt that the reduction of subsidies and a capacity glut have shaken out the solar industry's marginal players, the strongest companies continue to make progress towards grid parity. Though the company experienced tremendous difficulties between 2009 and 2011 due to collapsing panel prices and internal organizational problems, industry leader First Solar Inc. (NASDAQ:FSLR) continues to improve the conversion efficiency of its thin-film solar panels. According to their Q2 2011 corporate presentation, First Solar improved its conversion efficiency from 10.7 percent in 2008 to 11.8 percent in the first half of 2011, and is targeting further improvement to between 13.5 and 14.5 percent by 2014.[3] More importantly, First Solar has dropped its module manufacturing costs per watt from $1.08 in 2008 to $0.75 in 2011, and expects manufacturing costs to fall to between $0.63 and $0.52 per watt by 2014.[4]

While an in-depth review of all of the possibilities for alternative energy to mitigate the coming natural gas deliverability crisis is beyond the scope of this book, here are a few ways Americans can harness the incredible advances in the alternative power industry to reduce our consumption of natural gas.

a. Rooftop Solar Panels with Battery Storage

Until very recently, one of the biggest problems with photovoltaic (PV) solar power was the inability to store electricity when the sun is not shining. However, the combination of breakthroughs in lithium–ion batteries and improved solar panel performance is making the once prohibitive dream of cost effectively storing the sun's rays possible. While use of battery storage for residential installations is still very limited, it is certain to grow rapidly over the next few years. One company that is leading the charge into this arena is Solarcity Solar.[5] The largest installer of residential roof-mounted solar panels in the US and funded largely by Tesla Motors CEO Elon Musk, Solarcity is working to develop residential solar photovoltaic storage solutions that will allow homeowners to maximize the value of their solar panels.

b. Solar Hot Water Heaters

Though they may seem unsightly to some and are rarely seen on American rooftops, solar hot water heaters currently provide millions of gallons of hot water on a daily basis to millions of people all over the world. During the coming deliverability crisis, many residents of sunny and even not-so-sunny climes will trade in their natural gas and electric hot water heaters in favor of water heated by the sun.

c. Utility Scale Wind and Solar Projects with Battery Storage

Large cost reductions and performance improvements in wind turbines, solar technology and battery storage technology will soon destroy the perception of solar and wind as unreliable sources of electricity. Several wind farms with battery storage facilities designed to store wind-generated electricity have been built in the US in recent years, and more are likely to be built once power prices climb. For example, in September 2010, the Los Angeles Department of Water and Power (LADWP) announced that it planned to install a 5- to 10-megawatt storage facility from China's BYD Lighting at its wind power facility in the Tehachapi Mountains.[6] According to a story reported by website C/Net.com, the storage facility will help LADWP expand its delivery of renewable energy, as well as helping it comply with a recently passed California state bill that requires electric utilities to implement energy storage technologies by 2016.[7]

West Virginia's Laurel Mountain Wind Project is a wind generation plant comprised of 98 MW of wind generation and 32 MW of integrated battery-based energy storage. It is the largest installation of renewable energy with a dedicated battery storage facility in the US.[8] PJM Interconnection, the world's largest power marketer, will market electricity from Laurel Mountain.

It is not just your author who thinks battery storage systems will play a large role in the growth of the solar and wind industry. Terry Boston, President and CEO of PJM Interconnection, operator of Laurel Mountain, had this to say about the combination of battery storage and renewable power:

Energy storage technology is the silver bullet that helps resolve the variability in power demand.... Combining wind and solar with storage provides the greatest benefit to grid operations and has the potential to achieve the greatest economic value.[9]

The sunny climate of Spain has made the first commercial, utility-scale solar project with storage a reality. Torresol Energy's Gemasolar project, located in Seville, is the world's first commercial plant to generate and store electricity with solar technology, using a central tower receiver, a heliostat field and a molten-salt heat storage system.[10] Figure 20.01 shows the technology used in the Gemasolar project.

The 19.9-megawatt Gemasolar plant, which opened in April 2011, will supply power to twenty-five thousand homes around the clock, 365 days a year.[12] Due to the facility's molten storage tanks, which can provide up to fifteen hours of electrical generation without any solar feed, homes powered by Gemasolar will never again need electricity from a conventional power plant.[13] Torresol Energy is a joint venture between Spain's SENER Grupo de Ingenieria, S.A. (60-percent owner), a Spanish construction and engineering firm, and MASDAR (40-percent owner), an alternative energy company based in Abu Dubai that is constructing the world's first off-the-grid modern city.[14] In 2012 Torresol commissioned

Gemasolar: how it works

🗔1 Heliostats
Solar light is reflected by the heliostats towards the receiver, located on top of the tower.

🗖2 Tank 1
Molten salts, at 290ºC, are pumped from the cold molten salt tank to the receiver.

🏭3 Tower
Inside the receiver, molten salts are heated up to 565ºC before being stored in the hot molten salt tank.

🗖4 Tank 2
The hot molten salt tank keeps the energy accumulated in form of molten salts at very high temperature.

🖵5 Steam Generator
The hot molten salts are delivered to the steam generation system, where they transfer their heat to the water, reducing their temperature.

⊕6 Turbine
The heat transferred transforms the water into high pressure steam to move the turbine.

⊖7 Electric Generator
The turbine powers the electric generator producing electrical energy.

⚡8 Electrical Transformer
The electricity is delivered to a transformer to be injected into the distribution grid.

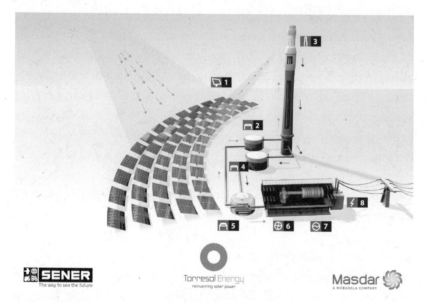

FIGURE 20.01. Gemasolar Technology Illustration.[11]

two other solar projects with storage capacity, adjacent 50-megawatt solar plants with storage in Cadiz, Spain, that generate enough electricity to supply forty-thousand homes.[15] The company is also pursuing opportunities to develop new solar generating facilities with storage in southern Europe, the Middle East, North Africa and the southwestern US.

d. Geothermal Heating and Cooling Systems

Though geothermal heating, cooling and hot water systems do not garner the attention other forms of alternative energy receive, they have been effective methods of heating hot water and controlling the climates of homes and commercial buildings for decades. By using fluid-filled pipes buried in a closed-loop system several feet below the earth's surface, which stays an average of 55 degrees in most climates all-year round, geothermal systems are able to collect heat and convert the earth's energy through an air exchange system to heat or cool a home or to heat water. While geothermal units use electricity to pump fluid through their closed-loop systems, they are usually significantly more efficient than conventional systems.

2. Increased Efficiency of the Electricity Grid

One of the dirtiest secrets in the utility business is that up to 77 percent of the energy contained in coal, uranium, water and natural gas is lost, either in running equipment, the conversion of the feedstock into electricity, or transmission to the point of consumption.[16] Clearly, America can significantly reduce its demand for natural gas from the electricity generation sector—which accounted for approximately 31 percent of all demand in 2011—by getting more out of the electricity we already generate.[17]

Here are two ways to make better use of America's existing electricity generation fleet using existing technology.

a. Improved Grid Management

Improved management of the electricity grid through the installation of smart meters in homes and businesses to allow for greater time-of-day metering will certainly reduce the electricity industry's demand for natural gas. By tracking consumption patterns, both residential and commercial consumers can make more informed decisions on how and when to consume electricity. However, the biggest increase in grid efficiency will come from battery storage.

One company that is delivering scalable grid management storage solutions is privately held GreenSmith Energy Management Systems.

Since it was founded in 2008, GreenSmith has installed grid management solutions using cutting-edge batteries, software and hardware to help utilities, university and commercial campuses and companies operating in remote locations increase efficiency.

b. Reduce Line Loss through Superconductors

While the US National Academy of Engineering has called America's electrical grid "the greatest engineering achievement of the 20th century," much of its installed infrastructure is reaching the end of its useful life and is in need of replacement.[18] With approximately 10 percent of all electricity generated lost through line transmission, upgrading the grid with superconductor wires and cables will have a very positive and immediate return on investment in most cases.[19] Several companies, such as AMSC (NASDAQ:AMSC) (formerly American Superconductor Corp) and France-based Nexans, have installed high-performance transmission products, inverters, converters and other offerings that will substantially reduce transmission loss in dozens of countries around the world.

3. Replace Natural Gas Peaking Units with Battery Storage

On the hottest days of the year—often a time of surging natural gas prices—utilities and independent power producers around the country dispatch inefficient, single-cycle natural gas power plants to meet surging electricity demand. I find this practice to be wasteful on three fronts. First, increasing demand for natural gas at times of high prices needlessly increases costs for utility companies and prices for consumers. Second, if we are going to burn natural gas to generate electricity, only efficient, combined-cycle plants should be used. As their name suggests, combined-cycle plants use natural gas to turn a turbine and capture the exhaust heat from the conversion process and use it to create steam to drive a second turbine for additional generation; single-cycle plants do not capture waste heat from the conversion process. Lastly, having capital tied up in single-cycle plants that may only be used a few days a year is very wasteful. A far better investment for utilities would be battery storage units that could increase grid efficiency year-round and be available on the hottest days of the year to supply incremental capacity.

4. Better Resource Management

For decades, better management of our endowment of natural gas has received virtually no attention from either policy makers in Washington or industry leaders. Rather, they have focused all of their attention in recent years on finding new uses for natural gas, due largely to the previously discussed 100-year supply myth. It's as if the country did not experience a natural gas crisis of epic proportions in the 1970s or live through the numerous price spikes of the early 2000s, which harmed our economy by driving manufacturing jobs to locales with lower natural gas prices, raising heating and electricity prices for consumers and damaging consumer confidence. Instead of finding new ways to use natural gas faster, I believe the US should focus on extending the life of its existing fields, encouraging exploration for new sources and adopting policies to facilitate the smarter development of known fields.

The Netherlands is a major gas-producing country that has taken a very enlightened approach to the development of its natural gas resources. The Dutch have adopted several policies to ensure their endowment of natural gas provides the maximum benefit to society, ensures security of supply and advances the country's economy.[20] One way the country is maximizing the benefits of its resources is through its Small Field Policy.

The Small Field Policy grew out of the Dutch government's desire to extend the life of its massive Groningen field, which was discovered in 1959 near Slochteren in the northeastern corner of the country.[21] The field contains approximately 39 tcf of the country's 50 tcf of total reserves; it is the largest field in Europe and the tenth largest in the world.[22] By the 1970s, the Dutch gas pipeline company Gasunie guaranteed the purchase of gas from small fields at market prices to allow for the use of Groningen as the "swing producer."[23] As you can see from Figure 20.2, virtually all of the country's production came from Groningen prior to the enactment of the Small Field Policy.

Despite the success of its Small Field Policy, by the mid-1990s the

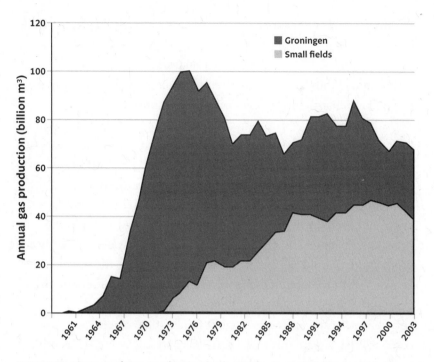

FIGURE 20.02. Annual Gas Production in the Netherlands in billion m³.[24]

Dutch government determined that additional resource management initiatives were needed to ensure long-term supply security. To achieve this end, in 1995 the government adopted the Third Energy Policy Paper that set an annual production ceiling of 2.85 tcf for the entire country, to ensure "a prudent use of the Groningen field."[25] Then, in 2003, after a downward adjustment of expected production from small fields and to extend the life of Groningen, the government passed the Natural Gas Law. This reduced the country's production ceiling to 2.68 tcf per year for years 2003 through 2007 with a further reduction to 2.47 tcf for years 2008 through 2013.[26]

So what can the US learn from the Dutch experience of resource management? While mandatory reductions in natural gas production may seem anti-competitive and un-American, the US actually has a long history of resource management. After the discovery of the massive oil fields in east Texas in 1930 and 1931, which drove the price of oil down from $1.10 per barrel prior to "Dad" Joiner's famous discovery on

Bradford Farm to only $0.15 per barrel less than two years later, a series of laws limiting production were enacted.[27] These laws largely remained in place until the US reached its peak oil production in the early 1970s. Establishing limits for US natural gas production will not be easy, since it will likely raise natural gas prices in the short term.

Another way to ensure America's natural gas resources are used for the maximum benefit of all stakeholders is to outlaw mineral leases that are less than five years in length. When an oil and gas company acquires land on which it plans to drill, it typically signs a three- or five-year lease that requires a one-time bonus payment to the leaseholder and the drilling of one well during the initial lease period. In addition to the one-time bonus payment, the leasor typically retains a royalty interest in any production. Should the leasee not drill a well during the initial period, he usually has the option of paying the landowner another bonus payment—which can amount to millions of dollars—to extend the lease for another few years, or he can attempt to negotiate a new lease. Often, a company that signs a three-year lease has a strong financial incentive to drill a well during the initial lease terms irrespective of prevailing natural gas prices. Once a producing well is drilled on a lease, it is said to be "held by production."

By extending lease terms to a minimum of five years, operators can take a more measured and intelligent approach to the development of prospective lands and not waste billions of dollars drilling uneconomic wells for the sole purpose of holding leases. For example, had prospective Haynesville Shale leases in northeast Louisiana been on five-year rather than three-year terms, operators would not have drilled hundreds of uneconomic wells costing billions of dollars between late 2008 and 2012. Additionally, requiring a minimum initial term of five years for oil and gas leases allows companies to perform the science necessary to efficiently develop pilot projects to test new areas for future development and to find the most effective and safe methods to fracture wells.

5. Rethinking the Importance of Nuclear Energy

While the Fukushima disaster cast a shadow over the entire nuclear industry in 2011 and caused many nations to question the role nuclear energy will play in the twenty-first century, there should be little doubt that it will be an important part of America's energy future. Nuclear energy is reliable, cheap, emits no greenhouse gases and the existing plants are largely paid for. During 2010, the nuclear power industry produced approximately 20 percent of US electricity from its 104 plants.[28] By comparison, during 2010 coal accounted for about 45 percent and natural gas around 24 percent of the nation's electricity supply.[29]

Although no new American nuclear plants have come online since 1996, when the last civilian nuclear reactor—TVA's Watts Barr Unit 1— was commissioned, output from the nuclear industry has grown significantly nonetheless. In fact, between 1996 and 2010, US nuclear power generation increased 19.6 percent through uprating and reduced downtime.[30] (It should be noted that the TVA's Watts Bar Unit 2, due to be completed between September and December 2015, is the only nuclear plant currently under construction in the US.[31]) In addition to uprating, the renewal of operating licenses by the Nuclear Regulatory Commission has extended the life of many US plants. By August of 2012, the NRC had extended the licenses of 73 reactors by 20 years and is considering license renewal applications for additional facilities.[32]

The increasingly stringent environmental regulations directed towards America's coal-fired power plants and the coming tightness in the natural gas market make a forward-looking approach to nuclear power a necessity. Industry executives and policy makers should be examining ways to extend the life of America's existing nuclear fleet where possible; build new, super-efficient reactors that set new standards for safety; and find a long-term storage solution for nuclear waste. Instead of natural gas being a "transition" or a "bridge fuel" to twenty-first-century energy independence, the coming natural gas deliverability crisis will perhaps make uranium the *new, new* bridge fuel.[33]

6. Adoption of Distributed Energy

Producing small amounts of electricity at or near the point of consumption—commonly referred to as distributed power—can substantially reduce electricity lost during transmission, change the balance of power in the utility industry and materially reduce America's consumption of natural gas. Distributed energy can take on several forms, such as rooftop or backyard solar arrays, natural gas-powered fuel cells for home and commercial buildings and even small-scale nuclear reactors. Before discussing the challenges facing the wide-scale adoption of distributed energy, let's review two of the more promising examples of distributed electricity generation.

a) Roof-Mounted Solar Panels
These are the most common form of distributed generation in America today by a large margin. As discussed previously, due to the coming spike in natural gas prices, increased panel efficiency and dropping costs, I expect roof-mounted solar panels to achieve grid parity in many markets in the next few years. Solar will continue to lead the distributed power revolution I see coming.

b) Efficient Fuel Cells
Fuel cells, which use very advanced materials to electrochemically convert natural gas into electricity at rates higher than conventional power plants, are likely to see wide-scale deployment over the next decade. One company that is leading the charge into fuel cell electricity generation is Silicon Valley's Bloom Energy. The company's ES-5700 Energy Server is a 200-kilowatt fuel cell based on planar solid oxide fuel cell technology first developed for NASA's Mars program.[34] It can convert natural gas into electricity at a greater than 50-percent efficiency ratio, on par with the most efficient combined-cycle turbines.[35] With each ES-5700—which the company calls "Bloom Boxes"—capable of generating enough electricity to power the baseload demand of 160 homes or one small office building, Bloom offers its customers the ability to scale its distributed generation capacity according to its needs.[36]

Funded by some of the biggest names in venture capital and headed by founder and CEO KR Sridhar, Ph.D, who worked on NASA's efforts to convert Martian atmospheric gases into oxygen for propulsion and life support, the company has already installed numerous Bloom Boxes for Fortune 500 clients.[37] Through a groundbreaking program with investment bank HSBC, several Bloom clients not only locked in below-grid electricity prices for ten years (HSBC has secured the gas supply through forward contracts) but also an electricity supply that is not subject to grid blackouts.[38]

It is not just solar and fuel cells that will help move the US away from the wastefulness of a centrally planned, grid-centric electricity industry. There is a movement underway for the adoption of very efficient, small-scale natural gas-fired power plants and even small-scale nuclear reactors to generate electricity near the point of consumption. For example, San Diego State University generates virtually all of its own electricity from a 14.4-megawatt cogeneration facility which includes two 5-megawatt, combined-cycle natural gas-fired turbines and two waste-heat boilers. Waste heat from the electricity generation process is recycled and used to generate more electricity if needed and for absorption chilling and heating of buildings. A feasibility study by the university completed in 2001 estimated the $22-million facility—commissioned the same year—would save the university $148 million over the next 30 years.[39]

While the first small-scale nuclear reactors probably won't be installed for several years—probably not before 2020—several companies are working hard to make distributed nuclear power a reality. Modular nuclear reactors can be used to produce electricity, to heat and cool campuses and to power remote towns, villages and factories, as well as to desalinate water. Construction giant Fluor Corporation became a significant player in the modular nuclear power business through its takeover of Corvallis, Oregon-based NuScale Power LLC in the fall of 2011. After running into financial difficulties, NuScale received a $3.5 million cash injection from Fluor, which plans to invest an additional $30 million into the company to advance its stand-alone 45 mega-watt distributed reactor towards commercialization.[40]

TerraPower LLC is a Bellevue, WA based start-up that is working on an innovative technology it refers to as a travelling wave reactor (TWR).[41] The company was formed in 2007 with the backing of Bill Gates and former Microsoft Chief Technology Officer Nathan Myhrvold and, according to CEO John Gilleland, has already spent tens of millions of dollars on research but would need billions more to build its first prototype travelling wave reactor.[42]

Charlotte, North Carolina-based engineering firm Babcock and Wilcox (NYSE: BWC) has partnered with privately held construction giant Bechtel Corporation to develop a commercially viable small-scale nuclear reactor. The alliance formed in 2009—dubbed "Generation mPower"—will focus on bringing to market Babcock and Wilcox's next generation of small-scale nuclear reactors.[43] In the fourth quarter of 2011, the partners began testing a prototype mPower reactor in Bedford, Virginia and collecting data.[44] Given BWC's more than fifty years of experience constructing nuclear plants throughout the world, Bechtel's more than sixty years of nuclear industry activity and the advanced stage of their prototype, I would expect Generation mPower to soon emerge as the recognized industry leader in the modular nuclear power industry.

The Fight to Squash Distributed Power

Despite all the positives of distributed power and laws such as FERC 888, known as the "open access rule," which requires open, non-discriminatory access to the transmission grid by third-party generators, regulated, incumbent private utilities clearly do not want competition from distributed power generators. In fact, there have been several cases where large, well-connected utilities have done everything possible to squash competition from smaller companies and private individuals selling excess electricity into the grid. For example, in 2010 San Francisco-based utility giant PG&E Corporation (NYSE:PCG) spent $44 million dollars in a campaign to make law ballot initiative Proposition 16, which would have reduced competition in electricity generation by limiting the ability of local governments to form power agencies to serve their communities.[45] Specifically, Proposition 16 would have required local municipalities to hold a referendum prior to setting up a new public power authority and receive a two-thirds majority vote in favor of creating the new authority, a historically difficult threshold to achieve.[46] Despite heavy lobbying efforts and campaigning by PG&E, the major opponent to Proposition 16—the Utility Reform Network— defeated it by spending a paltry $36,000.[47]

I believe the defeat of Proposition 16 is just the beginning of a distributed power revolution that will pit small producers such as homeowners selling excess power into the grid and small locally owned energy co-operatives against the large, regulated private utilities that continue to push for and receive rate increases. For example, according to the California Public Utility Commission, the average residential customer of PG&E saw his cost per watt of electricity rise from 10.7 cents per kilowatt hour (kWh) in 2000 to 15.7 cents per kWh in 2011—a 47 percent increase—while residential customers of San Diego Gas and Electric saw their rates rise from 11.3 cents per kWh in 2000 to 18.4 cents in 2011, a 63 percent increase.[48] Consumer outrage at further increases in electricity rates combined with falling costs and rising efficiency of alternate forms of generation will greatly accelerate the installation of millions of rooftop solar panels, fuel cells and other forms of distributed electricity generation across the state.

The combination of California's high electricity rates and several very disruptive blackouts in recent years should guarantee the state is ground zero for the nationwide distributed power revolution that I see coming.

CHAPTER 21

A Look at the Future

D UE LARGELY to the failure of natural gas prices to rebound along with those of other commodities such as oil and gold combined with the ads placed in mainstream media by energy companies or their trade associations such as the American Natural Gas Association (ANGA), many energy market observers are now convinced we have entered a "new era" of cheap natural gas for decades to come. Have we? There is substantial evidence that a number of factors combined to create a "perfect storm" that led to unsustainably low prices between 2009 and 2012. Here are the five main reasons natural gas, the one commodity left behind in a world that has been flooded with liquidity in recent years, will sharply rebound in price over the next couple of years.

1. Reduced Hedges Force Reduced Drilling

Though some savvy operators were able to protect their cash flow during this historic bear market by locking in higher prices through hedges, most firms no longer have a meaningful hedge book in place for 2013 and beyond. For example, a January 2012 article that appeared on the CommodityOnline website indicated that producers had not only hedged 8 percent less gas in 2012 than in 2011, but also that the gas was hedged at a lower price. According to the article, "Natural gas hedges for 2012 are clustered in the $5.52–5.67 range, which is about $0.42/MMBtu lower than hedges for 2011."[1]

With only 13 percent of their production hedged for 2013 and spot prices in late 2012 well below the break-even cost to bring on new production, shale gas operators have been massively reducing the number of rigs drilling for natural gas.[2] For example, EnCana Corporation, a Calgary-based company with substantial US operations in the Haynesville and Barnett shales, reduced its 2012 dry gas drilling budget by 37 percent.[3] The company also predicted that its gas production would fall 7.5 percent in 2012 due to the low price environment and rising costs.[4] In early 2012 another major natural gas player, Chesapeake Energy, announced that it planned to reduce its dry gas rig count by 68 percent.[5] One of the reasons behind Chesapeake's drastic reduction in natural gas spending, besides the company's financial difficulties (discussed earlier), was that it entered 2012 with hardly any of its gas production hedged, having closed out the vast majority of its natural gas hedge position in 2011.[6] The reduction in activity by EnCana, Chesapeake and a number of their peers has resulted in the Baker Hughes US natural gas-directed drill count falling from a high of 1,606 in 2008 to only 413 by November 2012, a drop of 74 percent.[7]

2. Held by Production (HBP) Drilling Now Over

One of the reasons for continued drilling between 2008 and 2012, despite low prices, was Held by Production (HBP) drilling—the need for operators to drill wells in order to retain their shale gas leases. Many of these leases had been signed during the shale land rush between 2005 and 2008, and to keep them in good standing, wells had to be drilled during the initial three- to five-year lease period. To avoid having to write off prospective acreage that cost billions of dollars to lease, dozens of shale gas operators drilled thousands of wells that had virtually no chance of making money. More importantly, HBP drilling temporarily depressed gas prices. Now that most prospective leases in the major shale plays have been drilled—there are still some undrilled, prospective leases in the Marcellus and Eagle Ford—it has largely come to an end. Companies are much more sensitive to the spot price of natural gas when making capital expenditure decisions now that the possibility of forfeiting millions of dollars of lease payments is greatly diminished.

3. Uncompleted Well Inventory Now Depleted

Because shale-focused companies were unable to schedule enough fracture stimulation crews while gas prices were rising during 2008 due to shortages of fracture crews, many built a significant inventory of uncompleted natural gas wells. In an effort to maintain production levels and keep their gas-directed spending to a minimum at times of low prices, companies such as Carrizo Oil and Gas, Devon Energy and Quicksilver Resources as well as many others have completed hundreds of wells in recent years from their inventory of wells that were drilled in 2008. While nearly every uptick in gas prices between 2009 and 2012 was met by completing previously drilled—but uncompleted wells—already in inventory, by late 2012, the cupboard was nearly bare.

4. Declining Productivity in New Shale Wells

One of the most overlooked reasons that the natural gas bear market of recent years will end shortly is the declining productivity of new shale wells. While this phenomena is not prevalent in all plays as of yet, it is certainly occurring in the Barnett, Fayetteville and Woodford (Arkoma Basin) shales. As I discussed in Chapter 16, Southwestern Energy, the leading operator in the Fayetteville Shale, is seeing less production from wells that have been online for 60 days despite drilling longer laterals and using more fracture stimulations per well. While shale operators have been able to compensate for declining prospect quality through drilling efficiencies such as pad drilling and increasing the number of fractures per well, lower average well productivity is a sure sign that a play's growth phase is nearly over.

5. Foreign Firms No Longer Overpaying to Participate in Shale Gas Plays

Far and away the most important factor in the brutal natural gas bear market between 2009 and 2012 was the influx of dollars from overseas operators eager to get into the shale business. Foreign companies have spent north of $100 billion dollars in the last six years to gain access to virtually every North American shale gas play. Two of the biggest drivers of this huge influx of foreign money were a desire to learn the shale gas business and a lack of investment opportunities elsewhere. Though many foreign firms have used their cash to take over US shale gas operators, tens of billions of dollars took the form of drilling carries that allowed their American partners to drill wells without regards to economics or prevailing natural gas prices. By early 2012, with the majority of drilling carries now complete, most American firms began to participate in future wells on a heads-up basis (i.e., both parties are required to pay a proportional share of drilling expenses). Instead of drilling additional wells at prices that would render most of them uneconomic, American firms have chosen to greatly reduce dry gas-directed drilling and have instead charged headstrong into unconventional oil plays.

While many promoters of shale gas are quick to point to the influx of foreign money as proof positive of the bright future of shale gas, I see things very differently. Similar to the situation when numerous European banks such as Dusseldorf's IKB Deutsche Industriebank and Dutch banking giant ABN Amro chronically overpaid for toxic AAA-rated, subprime mortgage-linked derivative products such as collateralized debt obligations (CDO)—IKB had to be bailed out in 2007 due to its failed US subprime-linked investments while ABN Amro was bailed out in 2009—I do not see the investments made by foreign oil and gas companies as a validation of the US shale gas business.[8] In fact, I see it as the opposite. Numerous US shale gas companies, helped by Wall Street firms positioned to collect transaction fees, vastly over-hyped the potential of US shale gas plays to get foreign firms to overpay for access to them. By mid-2012, numerous foreign companies such as the U.K.'s BG Group, BP and BHP—similar to their banking counterparts a few years earlier—had taken large write-downs on their shale gas investments.[9] As I discussed in Chapter 5, in a remarkable moment of candor at the height of the 2008 financial crisis, Aubrey McClendon of Chesapeake Energy confirmed to the world that an important part of the business model for his company was to sell land it bought cheaply for five to ten times the price it had paid.[10] Who better to sell to than foreign firms with deep pockets and little or no shale gas experience or expertise? Just as Wall Street firms dumped toxic mortgages and derivatives on foreign banks, US shale gas operators offloaded unproven acreage upon unwitting foreign companies and often signed deals that funded much of their drilling efforts between 2008 and 2012.

An Alternate Future Reality

Though many shale gas optimists paint a rosy picture in which America reduces or eliminates its need for foreign oil by converting much of its transportation fleet to natural gas, enjoys a cleaner environment by replacing coal-fired electricity generating plants with natural gas-fired plants and experiences a prolonged manufacturing renaissance thanks to our cheap new source of energy that puts hundreds of thousands of people back to work, I see a decidedly different future. While every oil and gas play is different, I see many parallels between the Austin Chalk play of Texas and Louisiana and modern shale gas plays. The Austin

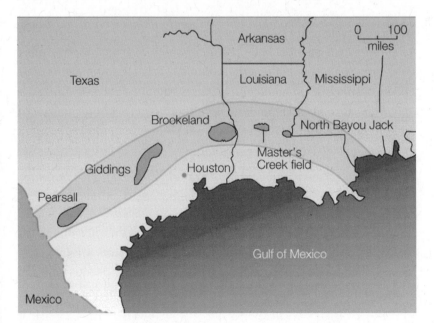

FIGURE 21.01. Map of Austin Chalk Play.[11]

Chalk play, located in a swath of limestone stretching from south-central Texas into Louisiana, was discovered in the early 1960s and sporadically developed using vertical wells until the late 1980s. Figure 21.01 identifies several of the largest Austin Chalk fields.

Early operators in the Austin Chalk found it very difficult to consistently produce commercial wells due to the depth of the formation and the cost of drilling; however, the application of horizontal drilling to the play in the early 1990s changed everything.[12] Dozens of companies rushed to lease acreage and drill thousands of horizontal wells, many of which were hydraulically fractured. The Austin Chalk was one of the first plays in the US to benefit from the implementation of horizontal drilling and hydraulic fracturing, and by 1996 the Giddings Field was the largest natural gas field in Texas, producing 296 bcf per annum.[13] It should be noted that the field was also the largest oil field in the state in 1993 with peak oil production that year of 32 million barrels.[14] Sound familiar? By the late 1990s, with limited remaining drilling locations and weak commodity prices, activity and production in the Austin Chalk collapsed. In 2011, gas production from Giddings Field was only 40 bcf, down 86.5 percent from its high in 1996.[15]

So what can the history of the Austin Chalk tell us about the future of modern shale plays? While the Austin Chalk is a limestone formation and has very different geological characteristics from shale formations, it provides a couple of important take-aways. First, relatively small sweet spots produced virtually all of the economic wells, similar to modern shale plays (I discussed the importance of sweet spots in shale plays extensively in Chapter 4). Though many operators attempted to replicate the success of Giddings and a handful of other fields in less prospective parts of the trend, few were successful. Second, the rapid rise in production from Giddings and other Austin Chalk fields due to the implementation of horizontal drilling and hydraulic fracturing is a great example of how these two technologies accelerate well decline rates and pull forward the date of a play's peak production. Today, only a handful of operators are active in the Austin Chalk and their activity consists largely of re-fracturing depleted wells and drilling a few dozen new wells every year.

History Set to Repeat

The drastic reduction in shale gas drilling that occurred in between 2010 and between 2010 and 2012 has set the stage for America's shale gas production growth to flatten for the first time in more than a decade. Does this mean that shale gas production is about to fall off a cliff? Absolutely not. However, flat shale gas production along with declines in offshore, CBM, tight gas, conventional production and Canadian imports, all at a time of rapidly rising demand, ensure that a deliverability crisis quietly lurks over the horizon. Once it arrives, many will wail, gnash their teeth, point fingers and claim, falsely, that it was impossible to see the crisis coming. In the 1970s, we were able to mitigate the gas crunch by building dozens of nuclear and hundreds of coal-fired power plants and by restricting use of gas in industrial boilers. These options are no longer available today, though demand is at an all-time high, so the upcoming deliverability crisis is certain to force individuals, the manufacturing sector and, once again, the electric utility industry to drastically scale back consumption. In other words, imagine a world where your home heating and electricity bills triple, food prices skyrocket due to increased costs for natural gas-based fertilizers and you lose your job because your employer cannot afford to pay his utility bills. But during the next crisis, unlike in the 1970s, there will be no easy way out.

> *"History does not repeat, but it does rhyme."*
> — Attributed to Mark Twain

Notes

Preface

1. David Lereah, *Why the Real Estate Boom Will Not Bust and How You Can Profit From It* (New York: Random House, 2005).
2. *The Economic Outlook: Before the Joint Economic Committee, U.S. Congress March 28, 2007* (Testimony of Chairman Ben S. Bernanke), accessed May 17, 2011, federalreserve.gov/newsevents/testimony/bernanke20071108a.htm, paragraph 6.
3. American Gas Association, "Snapshot of U.S. Natural Gas Consumption (2008)," accessed May 17, 2011, aga.org/our-issues/issuesummaries/Pages /SnapshotUSNaturalGas.aspx, paragraphs 1 and 4.
4. Tom Standing, "Alaska's Key to Oil Production—It's a Gas…" (Association for the Study of Peak Oil & Gas—USA, November 24, 2008), accessed May 17, 2011, aspousa.org/index.php/2008/11/alaskas-key-to-oil-production-its -a-gas., paragraph 13; US Energy Information Administration, "Alaska South Field Production of Crude Oil," release date November 29, 2012, accessed December 10, 2012, eia.gov/dnav/pet/hist/LeafHandler.ashx?n=PET &s=MCRFPAKS1&f=M.
5. US Energy Information Administration, "U.S. Natural Gas Imports (Million Cubic Feet)" accessed May 17, 2011, eia.gov/dnav/ng/hist/n9100us2A .htm, "U.S. Natural Gas Total Consumption (Million Cubic Feet)" accessed May 17, 2011, eia.gov/dnav/ng/hist/n9140us2A.htm.
6. US Energy Information Administration, "U.S. Natural Gas Marketed Production (Million Cubic Feet)" (1973), accessed May 17, 2011, eia.gov/dnav /ng/hist/n9050us2a.htm.

Chapter 1

1. EH.net, "Manufactured and Natural Gas Industry," (Economic History Association, February 1, 2010), accessed May 17, 2011, eh.net/encyclopedia /article/castaneda.gas.industry.us, paragraph 18.
2. Wikipedia.org, "Federal Trade Commission," accessed May 26, 2011, en.wiki pedia.org/wiki/Federal_Trade_Commission.
3. Report of the FTC to the US Senate, S. Doc. No. 92, 70th Cong., 1st Sess. 588-91 (1936), 28 as reported in endnote n2, Pierce, Richard J. "Reconsidering the Roles of Regulation and Competition in the Natural Gas Industry" (*Harvard Law Review*, December 1983, 97 Harv. L. Rev. 345).
4. Ibid.
5. Ibid.

6. Econ.com, "Appendix C Natural Gas Regulation in the United States," (*Economic Insight, Inc. February 1996*), accessed July 12, 2011, econ.com/appc.pdf.
7. Report of the FTC to the US Senate, S. Doc. No. 92, 70th Cong., 1st Sess. 588-91 (1936), 28 as reported in endnote n2, Pierce, Richard J. "Reconsidering the Roles of Regulation and Competition in the Natural Gas Industry" (*Harvard Law Review*, December 1983, 97 Harv. L. Rev. 345).
8. NaturalGas.org, "The History of Regulation—The Natural Gas Act of 1938," accessed May 17, 2011, naturalgas.org/regulation/history.asp#gasact1938.
9. US Energy Information Administration, "U.S. Natural Gas Wellhead Price (Dollars per Thousand Cubic Feet)," accessed May 17, 2011; average of 1950–54 prices, eia.doe.gov/dnav/ng/hist/n9190us3A.htm.
10. Rebecca L. Busby, Ed, Institute of Gas Technology, *Natural Gas in Nontechnical Language*, (Tulsa, OK: PennWell, 1999), 95.
11. NaturalGas.org, "History of Regulation—The Phillips Decision—Wellhead Price Regulation," accessed May 17, 2011, naturalgas.org/regulation/history .asp#phillips.
12. Ibid.
13. Ibid.
14. US Energy Information Administration, "U.S. Natural Gas Marketed Production (Million Cubic Feet)," accessed May 17, 2011, eia.gov/dnav/ng/hist /n9050us2a.htm.
15. Congressional Board, Office of Technology Assessment, "An Analysis of the Impacts of the Projected Natural Gas Curtailments for the Winter 1975–76," NTIS order #PB2-50623 (November 1975), fas.org/ota/reports/7502.pdf, 12.
16. Ibid.
17. Congressional Board, Office of Technology Assessment, "An Analysis of the Impacts of the Projected Natural Gas Curtailments for the Winter 1975–76," NTIS order #PB2-50623 (November 1975), fas.org/ota/reports/7502.pdf, 12; US Energy Information Administration, "U.S. Natural Gas Total Consumption (Million Cubic Feet)," accessed May 17, 2011, eia.gov/dnav/ng /hist/n9140us2a.htm.
18. Robert E. Wood, "Earnings Skid for Top Three Steel Makers," *LA Times*, April 29, 1970.
19. Arthur Everett, Associated Press, "Fierce Winter Cold Snap Causing Widespread Shortage of Oil, Gas," *The Lewiston (Maine) Daily Sun*, January 11, 1973, p. 6.
20. Reginald Stuart, "Factories Widen Search For Winter's Fuel Supply," *The New York Times*, November 17, 1975.
21. Reginald Stuart, "Ohio, Starved for Natural Gas, Strives to Adjust Its Way of Life," *The New York Times*, February 3, 1977.
22. Robert Sherrill, *The Oil Folies of 1970-1980* (New York: Anchor Press/ Doubleday, 1983), 294.

23. American Gas Association, "Our Country's Gas Supplies: What the Gas Industry is Doing to be Sure Your Home Has Enough Gas," *Life Magazine*, October 22, 1971, books.google.com/books?id=AEAEAAAAMBAJ&lpg =PA5&vq=American%20Gas%20association&pg=PA5#v=onepage&q&f =false.

24. US Energy Information Administration, "U.S. Natural Gas Marketed Production (Million Cubic Feet)," accessed May 17, 2011, eia.gov/dnav/ng/hist /n9050us2a.htm; US Energy Information Administration, "U.S. Natural Gas Wellhead Price (Dollars per Thousand Cubic Feet)," accessed May 17, 2011, eia.gov/dnav/ng/hist/n9190us3A.htm.

25. Erik Shuster, "Tracking New Coal Fired Power Plants" (data update 1/13/ 2012), presentation US Department of Energy, National Energy Technology Laboratory, January 13, 2012, 10.

26. US Energy Information Administration, "Electricity Net Generation: Total (All Sectors), Selected Years, 1949–2009," accessed May 17, 2011, eia.doe.gov /emeu/aer/pdf/pages/sec8_8.pdf.

27. US Energy Information Administration, "Table 7.2a Electricity Net Generation: Total (All Sectors)," accessed March 7, 2012, eia.gov/electricity/data .cfm#generation.

28. President Richard Nixon to Congress, April 18, 1973. The American Presidency Project, accessed May 17, 2011, presidency.ucsb.edu/ws/index.php ?pid=3817&st=&st1#axzz1MjSbsAaV, paragraph 35.

29. NaturalGas.org, "The History of Regulation—The Natural Gas Act of 1938," accessed May 17, 2011 naturalgas.org/regulation/history.asp#gasact1938; Robert R. Nordhaus, "Producer Regulation and the Natural Policy Act of 1978," *Natural Resources Journal*, October 1979, Vol. 19, lawlibrary.unm.edu /nrj/19/4/04_nordhaus_producer.pdf, p. 835.

30. Federal Power Commissioner Rush Moody, Jr. to President Gerald Ford, March 7, 1975, The American Presidency Project, accessed May 17, 2011, presidency.ucsb.edu/ws/index.php?pid=4769&st=natural+gas&st1#axzz1 MjSbsAaV, paragraph 3.

31. FTC, Opinion 770-A, cited in Joseph P. Mulholland, Federal Trade Commission, "Economic Structure and Behavior in the Natural Gas Production Industry," February 1979.

32. President Jimmy Carter to the Speaker of the House and the President of the Senate, January 26, 1977, The American Presidency Project, accessed May 17, 2011, presidency.ucsb.edu/ws/index.php?pid=7156#ixzz1MjZR2 CKZ, paragraph 9.

33. President Jimmy Carter remarks on the Emergency Natural Gas Act of 1977—remarks on signing S.474 and Related Documents, February 2, 1977, The American Presidency Project, accessed May 17, 2011, presidency.ucsb .edu/ws/index.php?pid=7422#axzz1ISpT64e1, paragraph 4.

34. Department of the Treasury, Bureau of the Mint, "Mint Facilities to Curtail Use of Natural Gas," press release, January 28, 1977, accessed May 17, 2011, usmint.gov/historianscorner/docs/pr514.pdf.

35. Rebecca L. Busby, Ed, Institute of Gas Technology, *Natural Gas in Nontechnical Language*, (Tulsa, OK: PennWell, 1999), 98.

36. Ibid; Institute for 21st Century Energy, "1980s: Free Markets and the Decline of OPEC," accessed May 17, 2011, energyxxi.org/reports/1980s.pdf, 28.

37. NaturalGas.org, "The History of Regulation, The Natural Gas Policy Act of 1978," Accessed, May 17, 2011 naturalgas.org/regulation/history.asp#gas act1978.

38. Ibid.

Chapter 2

1. US Energy Information Administration, "U.S. Natural Gas Wellhead Price (Dollars per Thousand Cubic Feet)," accessed May 17, 2011, eia.gov/dnav/ng /hist/n9190us3a.htm; US Energy Information Administration, "U.S. Natural Gas Marketed Production (Million Cubic Feet)," eia.gov/dnav/ng/hist /n9050us2a.htm.

2. Congressional Budget Office, "Natural Gas Price Decontrol: A Comparison of Two Bills" (Congress of the United States, November 1983), accessed May 17, 2011, cbo.gov/doc.cfm?index=5076, 3.

3. Ibid, 3.

4. ibid, 4.

5. Environmental Defense Fund, "Clean Air Act Timeline—A Short History of Key Moments in One of the Most Effective Public Health Campaigns in U.S. History," accessed May 17, 2011 edf.org/documents/2695_cleanairact .htm.

6. US Energy Information Administration, "Consumption for Electricity Generation by Energy Source: Total (All Sectors)," accessed May 19, 2011, eia.doe.gov/emeu/aer/txt/ptb0804a.html.

7. Ibid.

8. US Energy Information Administration, "Natural Gas Consumption by Sector, Selected Years 1949-2011 (Billion Cubic Feet)," Accessed December 10, 2012,, eia.gov/totalenergy/data/annual/pdf/sec6_13.pdf; US Department of Commerce, Bureau of Economic Analysis, "National Income and Product Accounts Table—Percent Change From Preceding Period in Real Gross Domestic Product," Accessed May 19, 2011, Years 1978-1987 selected, bea.gov/national/nipaweb/TableView.asp?SelectedTable=1&ViewSeries =NO&Java=no&Request3Place=N&3Place=N&FromView=YES&Freq= Year&FirstYear=1978&LastYear=1987&3Place=N&Update=Update&Java Box=no#Mid.

Chapter 3

1. NaturalGas.org, "The History of Regulation, The Move towards Deregulation," accessed, May 17, 2011 naturalgas.org/regulation/history.asp#dereg.
2. Ibid.
3. NaturalGas.org, "The History of Regulation, FERC Order No. 436," accessed May 17, 2011, naturalgas.org/regulation/history.asp#ferc436.
4. Ibid.
5. NaturalGas.org, "The History of Regulation, The Natural Gas Wellhead Decontrol Act of 1989," accessed May 17, 2011, naturalgas.org/regulation /history.asp#wellhead.
6. NaturalGas.org, "The History of Regulation, FERC Order No. 636," accessed May 17, 2011, naturalgas.org/regulation/history.asp#ferc636.
7. US Energy Information Administration, "U.S. Natural Gas Marketed Production (Million Cubic Feet)," accessed May 19, 2011, eia.gov/dnav/ng/hist /n9050us2a.htm; US Energy Information Administration, "U.S. Natural Gas Pipeline Imports From Canada (Million Cubic Feet)," accessed May 19, 2011, eia.gov/dnav/ng/hist/n9102cn2a.htm; US Energy Information Administration, "U.S. Natural Gas Wellhead Price (Dollars per Thousand Cubic Feet)," accessed May 19, 2011, eia.gov/dnav/ng/hist/n9190us3a.htm; Baker Hughes Incorporated, "North America Rotary Rig Count Current Week Data/ US Oil/Gas Split," accessed May 19, 2011, investor.shareholder .com/bhi/rig_counts/rc_index.cfm; US Energy Information Administration, "U.S. Natural Gas Number of Gas and Gas Condensate Wells (Number of Elements)," accessed May 19, 2011, eia.gov/dnav/ng/hist/na1170_nus _8a.htm.
8. US Energy Information Administration, "Repeal of the Powerplant and Industrial Fuel Use Act (1987)," accessed May 19, 2011, eia.doe.gov/oil_gas /natural_gas/analysis_publications/ngmajorleg/repeal.html.
9. US Environmental Protection Agency, "Overview - The Clean Air Act Amendments of 1990," accessed May 19, 2011, epa.gov/oar/caa/caaa_over view.html.
10. US Environmental Protection Agency, "Reducing Acid Rain," accessed May 19, 2011, epa.gov/air/peg/acidrain.html.
11. Robert Bryce, *Pipe Dreams* (New York, Public Affairs, 2002), 54.
12. US Energy Information Administration, "Annual Energy Review 2009: Natural Gas Consumption by Sector, Selected Years, 1949-2009 (Billion Cubic Feet)," accessed May 19, 2011, eia.doe.gov/emeu/aer/pdf/pages/sec6_13 .pdf.
13. US Energy Information Administration, "Henry Hub Gulf Coast Natural Gas Spot Price (Dollars/Mil. BTUs)," accessed May 19, 2011, eia.gov/dnav /ng/hist/rngwhhdd.htm.

14. US Energy Information Administration, "U.S. Natural Gas Wellhead Price (Dollars per Thousand Cubic Feet) — Annual Data," Release 11/30/2012, accessed December 11, 2012, eia.gov/dnav/ng/hist/n9190us3m.htm.

Chapter 4

1. US Energy Information Administration, "U.S. Natural Gas Wellhead Price (Dollars per Thousand Cubic Feet)," accessed June 7, 2011, eia.gov/dnav/ng/hist/n9190us3a.html.
2. Statistics Canada. Table 131-0001—Supply and Disposition of Natural Gas, Monthly (Cubic Metres), CANSIM (database), accessed May 6, 2011, www5.statcan.gc.ca/cansim/a01?lang=eng.
3. Baker Hughes, "North America Rotary Rig Counts—Tab Canadian Oil and Gas Split," accessed June 7, 2011, investor.shareholder.com/bhi/rig_counts/rc_index.cfm.
4. US Energy Information Administration, "U.S. Natural Gas Pipeline Imports From Canada (Million Cubic Feet)," accessed June 7, 2011, eia.gov/dnav/ng/hist/n9102cn2a.htm.
5. US Energy Information Administration, "U.S. Natural Gas Marketed Production (Million Cubic Feet)," accessed May 17, 2011, eia.gov/dnav/ng/hist/n9050us2a.htm.
6. Baker Hughes, "North America Rotary Rig Counts—Tab Canadian Oil and Gas Split," accessed June 7, 2011, investor.shareholder.com/bhi/rig_counts/rc_index.cfm.
7. Statistics Canada. Table 131-0001—Supply and Disposition of Natural Gas, Monthly (Cubic Metres), CANSIM (database), accessed May 6, 2011, 5.statcan.gc.ca/cansim/a01?lang=eng; Natural Resources Canada, "Canadian Natural Gas" Monthly Update October 2012, accessed December 11, 2012, .nrcan.gc.ca/energy/sources/natural-gas/monthly-market-update/2258.
8. Statistics Canada. Table 131-0001—Supply and Disposition of Natural Gas, Monthly (Cubic Metres), CANSIM (database), accessed May 6, 2011, 5.statcan.gc.ca/cansim/a01?lang=eng.
9. Baker Hughes, "North America Rotary Rig Counts—Tab Canadian Oil and Gas Split," accessed June 7, 2011, investor.shareholder.com/bhi/rig_counts/rc_index.cfm.
10. US Energy Information Administration, "U.S. Natural Gas Pipeline Imports From Canada (Million Cubic Feet)," accessed June 7, 2011 eia.gov/dnav/ng/hist/n9102cn2a.htm.
11. US Energy Information Administration, "U.S. Natural Gas Marketed Production (Million Cubic Feet)," accessed May 17, 2011, eia.gov/dnav/ng/hist/n9050us2a.htm.
12. Ibid.
13. Baker Hughes, "North America Rotary Rig Counts—Tab US Oil and Gas

Split," accessed June 7, 2011, investor.shareholder.com/bhi/rig_counts/rc_index.cfm.

14. US Energy Information Administration, "U.S. Natural Gas Wellhead Price (Dollars per Thousand Cubic Feet)," accessed May 17, 2011, eia.gov/dnav/ng/hist/n9190us3a.htm.

15. US Energy Information Administration, "U.S. Natural Gas Number of Gas and Gas Condensate Wells (Number of Elements)," accessed June 7, 2011, eia.gov/dnav/ng/hist/na1170_nus_8a.htm.

16. US Energy Information Administration, "Federal Offshore—Gulf of Mexico Natural Gas Marketed Production (Million Cubic Feet)," accessed June 8, 2011, eia.gov/dnav/ng/hist/n9050fx2a.htm.

17. US Energy Information Administration, "Natural Gas 2006 Year-in-Review," accessed June 8, 2011, eia.gov/pub/oil_gas/natural_gas/feature_articles/2007/ngyir2006/ngyir2006.pdf.

18. Natural Gas Intelligence, "North American LNG Import Terminals," accessed June 8, 2011, intelligencepress.com/features/lng/.

19. FERC.gov, "North American LNG Import Terminals—Existing as of December 5, 2012," accessed December 8, 2012, ferc.gov/industries/gas/indus-act/lng/LNG-existing.pdf.

20. Ibid.

21. FERC.gov, "North American LNG Import/Export Terminals—Approved," accessed July 17, 2012, ferc.gov/industries/gas/indus-act/lng/LNG-approved.pdf.

22. US Energy Information Administration, "U.S. Liquefied Natural Gas Imports (Million Cubic Feet)," accessed June 17, 2011, eia.gov/dnav/ng/hist/n9103us2a.htm.

23. US Energy Information Administration, "U.S. Natural Gas Wellhead Price (Dollars per Thousand Cubic Feet)," accessed June 7, 2011, eia.gov/dnav/ng/hist/n9190us3a.htm.

24. US Energy Information Administration, "U.S. Natural Gas Total Consumption (Million Cubic Feet)," accessed June 17, 2011, eia.gov/dnav/ng/hist/n9140us2a.htm.

25. Jon Rigby, CFA, "Global Nat Gas: When Will the Flood Ebb?" (UBS Investment Research Q-Seriesâ: Global Oil and Gas, February 26, 2010).

26. Ibid.

27. US Energy Information Administration, "Country Analysis Brief: Kuwait," accessed June 22, 2011, eia.gov/cabs/Kuwait/Full.html.

28. bp.com, "BP Statistical Review of World Energy June 2012," accessed August 23, 2012, bp.com/assets/bp_internet/globalbp/globalbp_uk_english/reports_and_publications/statistical_energy_review_2011/STAGING/local_assets/pdf/statistical_review_of_world_energy_full_report_2012.pdf, p. 28.

29. Ibid; US Energy Information Administration, "Country Analysis Brief: Qatar," accessed June 22, 2011, eia.gov/countries/cab.cfm?fips=QA; US Department of Energy, "Qatar Accounts for a Growing Share of LNG Exports," accessed June 20, 2011, eia.gov/todayinenergy/detail.cfm?id=50; Reuters, "Qatar to Hit Full LNG Export Capacity End 2011," June 5, 2011, accessed June 20, 2011, arabnews.com/node/379711448755.

30. US Energy Information Administration, "Country Analysis Brief: Qatar," accessed June 22, 2011, eia.gov/countries/cab.cfm?fips=QA.

31. *Oil and Gas Journal*, quoted in US Energy Information Administration, "Country Analysis Brief: Qatar," accessed June 22, 2011, eia.gov/countries /cab.cfm?fips=QA.

32. Ibid.

33. Matt Simmons interviewed in Dave Cohen, "Questions About the World's Biggest Natural Gas Field," (*The Oil Drum*, June 9, 2006, 6:41 pm), accessed June 22, 2011, theoildrum.com/story/2006/6/8/155013/7696.

34. Department of Energy & Climate Change, "Natural Gas Imports and Exports (Dukes 4.3)," accessed June 27, 2011, decc.gov.uk/en/content/cms /statistics/energy_stats/source/gas/gas.aspx; Department of Energy & Climate Change, "Natural gas supply and consumption (ET 4.1), accessed June 27, 2011, decc.gov.uk/en/content/cms/statistics/energy_stats/source /gas/gas.aspx.

35. Department of Energy & Climate Change, "Power Stations in the United Kingdom, May 2010," accessed June 27, 2011, decc.gov.uk/en/content/cms /statistics/energy_stats/source/electricity/electricity.aspx; Department of Energy & Climate Change, "Fuel Used in Electricity Generation and Electricity Supplied (ET 5.1)," accessed June 27, 2011, decc.gov.uk/en/content /cms/statistics/energy_stats/source/electricity/electricity.aspx.

36. Department of Energy & Climate Change, "Natural Gas and Colliery Methane Production and Consumption, 1970 to 2009," accessed June 27, 2011, decc.gov.uk/en/content/cms/statistics/energy_stats/source/gas/gas.aspx; US Energy Information Administration, "North Sea, Europe," republished by eoearth.org, accessed August 29, 2011, www.eoearth.org/article/North _Sea,_Europe, paragraph 3.1.2.

37. bp.com, "BP Statistical Review of World Energy, June 2011," bp.com/assets /bp_internet/globalbp/globalbp_uk_english/reports_and_publications /statistical_energy_review_2011/STAGING/local_assets/pdf/statistical _review_of_world_energy_full_report_2011.pdf, p. 22.

38. bp.com, "BP Statistical Review of World Energy, June 2012," accessed August 23, 2012, bp.com/assets/bp_internet/globalbp/globalbp_uk_english /reports_and_publications/statistical_energy_review_2011/STAGING /local_assets/pdf/statistical_review_of_world_energy_full_report_2012 .pdf, p. 22. BBC News, " Fresh Alert over UK Gas Supplies—National Grid

Has Issued its Latest "Balancing Alert" on Gas Supplies," accessed June 27, 2011, news.bbc.co.uk/2/hi/business/8452805.stm.

39. US Energy Information Administration, "What is Shale Gas and Why is it Important?," accessed June 27, 2011, eia.doe.gov/energy_in_brief/about _shale_gas.cfm.

40. Eileen and Gary Lash, "Kicking Down the Well," (The SUNY Fredonia Shale Research Institute, The Early History of Natural Gas), accessed June 27, 2011, aapg.org/explorer/2011/09sep/natgashisto911.cfm.

41. Devon Energy, Newsroom—News Release, "Devon Energy to Acquire Mitchell Energy for $3.5 Billion," accessed June 27, 2011, devonenergy.com /NEWSROOM/Pages/NewsRelease.aspx?id=200032.

42. AAPG.org, "A 17-Year Overnight Sensation, Barnett Shale Play Going Strong," accessed August 29, 2011, aapg.org/explorer/2005/05may/barnett _shale.cfm.

43. Texas Railroad Commission, "Statewide Production Data Query, Field: Newark East, Barnett Shale, 2008–2010," accessed June 27, 2011, webapps.rrc .state.tx.us/PDQ/changePeriodAction.do.

44. State of Arkansas Oil and Gas Commission, "B-43 Field Well MCF (Totals through 5/31/11 unless noted 8/25/11)," accessed August 29, 2011, aogc.state .ar.us/Fayprodinfo.htm.

45. US Energy Information Administration, "Lower 48 States Shale Plays," (Energy Information Administration, May 9, 2011), accessed June 27, 2011, eia.gov/oil_gas/rpd/shale_gas.jpg.

46. US Energy Information Administration, "Shale Gas Production (Billion Cubic Feet)," accessed June 27, 2011, eia.gov/dnav/ng/ng_prod_shalegas _s1_a.htm.

47. Texas Railroad Commission, "Statewide Production Data Query, Field: Newark East, Barnett Shale, 2008–2010," accessed June 27, 2011, webapps.rrc .state.tx.us/PDQ/generalReportAction.do.

48. Texas Railroad Commission, "Barnett Shale Information—Updated July 20, 2012" and "Newark, East (Barnett Shale) Statistics," accessed August 27, 2012, rrc.state.tx.us/barnettshale/index.php.

49. Arkansas Oil and Gas Commission, "Fayetteville Shale Gas Sales Information, Lifetime Totals—All Counties 3/31/11," accessed June 28, 2011, aogc .state.ar.us/Fayprodinfo.htm.

50. Arthur Berman, "ExxonMobil's Acquisition of XTO Energy: The Fallacy of the Manufacturing Model in Shale Plays" (*World Oil*, February 22, 2010), accessed May 20, 2011, theoildrum.com/node/6229.

51. Arthur Berman, Labyrinth Consulting Services, Inc., "Shale Gas—Abundance or Mirage, Why the Marcellus Shale Will Disappoint Expectations," (ASPO USA 2010 World Oil Conference, Washington, D.C. October 8, 2010).

52. Barnett Shale Maps—Barnett Shale Maps, Specific Source unknown, accessed June 28, 2011, blumtexas.blogspot.com/2007/05/blog-post_190.html and 3.bp.blogspot.com/_tiLoSTplBIU/RstHX5S230I/AAAAAAAAAc4/ZE4J88LMdMg/s1600-h/bsmap51.jpg.

53. Aubrey McClendon, Bloomberg News, October 14, 2009, as quoted in Arthur Berman, "Shale Gas—Abundance or Mirage? Why The Marcellus Shale Will Disappoint Expectations," theoildrum.com, October 28, 2010 - 10:20am, accessed June 28, 2011, theoildrum.com/node/7075.

54. Arthur Berman, Labyrinth Consulting Services, Inc., "Natural Gas Supply: Not as Great or as Inexpensive as Commonly Believed," American Association of Petroleum Geologists Conference, Houston, Texas, April 12, 2011, accessed June 28, 2011, searchanddiscovery.com/documents/2011/700 97berman/ndx_berman.pdf.

55. Ibid.

56. Arthur Berman, Labyrinth Consulting Services, Inc., "Will The Plays Be Commercial? Impact on Natural Gas Price" (Middlefield Capital Presentation, July 2010), slide 8.

57. Fetete.com, F.A.S.T. CBM™ Coalbed Methane Reservoir Analysis, "Decline Curve Analysis—Exponential, Hyperbolic, Harmonic Declines Illustration," accessed June 28, 2011, fekete.com/software/cbm/media/webhelp/c-te-techniques.htm.

58. M. J. Fetkovich, E. J. Fetkovich and M.D. Fetkovich, Phillips Petroleum Co., "Useful Concepts for Decline-Curve Forecasting, Reserve Estimation, and Analysis" (Society of Petroleum Engineers Reservoir Engineering, February 1996), Number 00028628.

59. Chesapeake Energy, "2010 Institutional Investor and Analyst Meeting," Oklahoma City, OK, October 13, 2010, quoted in Ian Urbina, "Drilling Down—Insiders Sound an Alarm Amid a Natural Gas Rush," "Drilling Down, Documents: Leaked Industry E-mails and Reports," accessed November 13, 2011, nytimes.com/interactive/us/natural-gas-drilling-down -documents-4.html., p. 202.

60. Vince White, Senior Vice President—Investor Relations, Devon Energy Corporation, "Interview with Bill Powers," May 24, 2011, 1:58 pm, Recorded with permission, duration 37:59. iTunes Voice Memo format.

61. Vince White, Senior Vice President—Investor Relations, Devon Energy Corporation, "Interview with Bill Powers," May 24, 2011, 1:58 pm, Recorded with permission, duration 37:59. iTunes Voice Memo format. .

62. Ben P. Dell and Noam Lockshin, "Bernstein E&Ps: The Death Throes of the Barnett Shale? Downgrading Devon to Market-Perform," Bernstein Research, May 13, 2010.

63. Ben Dell, email message to author, June 20, 2011.

64. Aubrey McClendon, CEO, Chesapeake Energy Corporation, "Chesapeake

Energy Corporation (CHK) Q2 2009 Earnings Call, Question-and-Answer-Session," August 4, 2009 9:00 AM ET seekingalpha.com/article /153691-chesapeake-energy-corporation-q2-2009-earnings-call-transcript and seekingalpha.com/article/153691-chesapeake-energy-corporation-q2 -2009-earnings-call-transcript?part=qanda.

65. Arthur Berman, Labyrinth Consulting Services, Inc., "Will The Plays Be Commercial? Impact on Natural Gas Price" (Middlefield Capital Presentation, July 2010), slide 10.

66. Business Wire, "Chesapeake Energy Corporation Comments on Inaccurate and Misleading *New York Times* Article," June 27, 2011 5:52 am, accessed September 1, 2011, chk.com/news/articles/Pages/1579995.aspx.

67. Chesapeake Energy, 2011 Form 10-K, accessed August 22, 2012, phx.corpor ate-ir.net/External.File?item=UGFyZW50SUQ9MTMyNDcofENoaWxk SUQ9LTF8VHlwZToz&t=1, p. 16.

68. Range Resources, 2010 Annual Report, accessed September 4, 2011, phx .corporate-ir.net/phoenix.zhtml?c=101196&p=irol-reportsAnnual, page 43.

69. "US Shale and Australian Nickel Asset Review," BHP Billiton Limited, accessed August 28, 2012, bhpbilliton.com/home/investors/news/Pages /Articles/US-Shale-and-Australian-Nickel-Asset-Review.aspx.

70. Exco Resources, Inc., "Exco Resources, Inc. Reports Second Quarter 2012 Results," news release, July 31, 2012, media.corporate-ir.net/media_files/irol /19/195412/XCO_Q2_2012_PressRelease.pdf.

71. Chesapeake Energy Corporation, "Chesapeake Energy Corporation Reports Financial and Operational Results for the 2012 Second Quarter," news release, August 6, 2012 11:24 am, chk.com/news/articles/Pages/1722883 .aspx.

72. George H. "Bud" Lawrence, "Turnaround, Stories of: Behind the Scenes Political Washington, A Great American Industry and Fun Along the Way," (Stillwater, OK: New Forums Press, Inc., 2007), p. 94.

73. John Curtis, PhD, Chairman of the Potential Gas Committee, "Interview with Bill Powers," June 6, 2011, recorded with permission, duration 1:19:46. iTunes Voice Memo format.

74. Ibid.

75. US Geological Survey, "U.S. Geological Survey 2002 Petroleum Resource Assessment of the National Petroleum Reserve in Alaska (NPRA)," (Fact Sheet 045–02, 2002), accessed July 2, 2011, pubs.usgs.gov/fs/2002/fs045-02/.

76. US Geological Survey, "National Oil and Gas Assessment Project, 2010 Updated Assessment of Undiscovered Oil and Gas Resources of the National Petroleum Reserve in Alaska (NPRA)," (Fact Sheet 2010–3102, October 2010), accessed July 2, 2011 pubs.usgs.gov/fs/2010/3102/.

77. Potential Gas Committee, "Potential Supply of Natural Gas in the United States, Report of the Potential Gas Committee (December 31, 2010),"

(Washington, DC, April 27, 2011), accessed July 2, 2011, potentialgas.org/ and PGC Press Conf 2011 slides.pdf linked from site, slide 8.

78. US Energy Information Administration, "International Energy Statistics— Natural Gas Proved Reserves," accessed July 2, 2011, eia.gov/cfapps/ipdb project/IEDIndex3.cfm?tid=3&pid=3&aid=6.

79. Potential Gas Committee, "Potential Supply of Natural Gas in the United States, Report of the Potential Gas Committee (December 31, 2010)," (Washington, DC, April 27, 2011), accessed July 2, 2011, potentialgas.org/ and PGC Press Conf 2011 slides.pdf linked from site, slide 8.

80. Kathy Shirley, "Country Has 'Abundant' Potential Report Tracks U.S. Gas Reserves," (Explorer), accessed June 8, 2011, aapg.org/explorer/2001/06jun /gas_update.cfm.

81. Gregory S. McRae and Carolyn Ruppel et al, "The Future of Natural Gas, an Interdisciplinary MIT Study" (MIT, June 2011), accessed July 25, 2011, web.mit.edu/mitei/research/studies/natural-gas-2011.shtml, page iii; American Clean Skies Foundation, "About" page, accessed July 25, 2011, cleanskies.org/about/; Navigant Consulting Inc., "North American Natural Gas Supply Assessment, Prepared For American Clean Skies Foundation," July 4, 2008, accessed July 25, 2011, cleanskies.org/pdf/navigant-natural-gas -supply-0708.pdf.

82. ExxonMobil, https://www.youtube.com/watch?v=XDfVycbnaBc.

83. Exxon Mobil Corp, "10K Annual report pursuant to section 13 and 15(d), Filed on 02/25/2011, Filed Period 12/31/2010, accessed July 25, 2011, ir.exxon mobil.com/phoenix.zhtml?c=115024&p=irol-SECText&TEXT=aHR0cDo vL2FwaS5oZW5rrd2l6YXJkLmNvbS9maWWxpbmcueG1sP2lwYWdlPTcoM zgxNjkmRFNFUT0wJlNFUT0wJlNRREVVTQz1TRUNUUSU9OX0VOVE lSRSZzdWJzaWQ9NTc%3d, p. 10.

Chapter 5

1. "What's Cooking with Gas: the Role of Natural Gas in Energy Indepen- dence and Global Warming Solutions," before the Select Committee on Energy Independence and Global Warming, 110th Congress (July 30, 2006) (statement of Aubrey McClendon, CEO of Chesapeake Energy), accessed July 2, 2011, globalwarming.house.gov/pubs/archives_110?id=0051#main _content and transcript: 110-46_2008-07-30.pdf.

2. Ibid., p. 78.

3. Ibid., p. 80.

4. American Gas Association, "Our Country's Gas Supplies: What the Gas In- dustry is Doing to be Sure Your Home has Enough Gas," *Life* magazine, Oc- tober 22, 1971, books.google.com/books?id=AEAEAAAAMBAJ&lpg=PA 5&vq=American%20Gas%20association&pg=PA5#v=onepage&q&f=false.

5. Aubrey K. McClendon, "Written Testimony of Aubrey K. McClendon, Chairman and CEO of Chesapeake Energy Corporation and Chairman of the American Clean Skies Foundation, before The Select Committee on Energy Independence & Global Warming—July 30, 2008," accessed July 2, 2011, republicans.globalwarming.sensenbrenner.house.gov/media/file /PDFs/Hearings/2008/073008/CookingWithGas/Aubrey/McClendon Testimony.pdf, p. 1.

6. Author estimate based on US Department of Energy, "Natural Gas Wellhead Value and Marketed Production," 2010 Marketed Production, accessed December 15, 2012, eia.gov/dnav/ng/ng_prod_whv_dcu_nus_a.htm.

7. Aubrey McClendon, CEO, Chesapeake Energy Corporation, "Chesapeake Energy Corporation (CHK) *Chesapeake Energy Corp Media Conference Call,*" July 11, 2011, recorded, duration 28:38, iTunes Voice Memo format.

8. Ibid.

9. Aubrey McClendon, interview with Leslie Stahl, Shalionaires (CBS "60 Minutes," November 14, 2010), FLV video, 13:25, youtube.com/watch?v=Vr 6b-WzIcyo.

10. bp.com, "BP Statistical Review of World Energy June 2011," accessed June 27, 2011, bp.com/sectionbodycopy.do?categoryId=7500&contentId =7068481, p. 6.

11. Aubrey McClendon, "Chesapeake Energy Corporation (CHK) Q3 2008 Business Update Call," October 15, 2008 2:30 PM ET, transcript, quoted in Seekingalpha.com, accessed July 26, 2011, seekingalpha.com/article /100644-chesapeake-energy-corporation-q3-2008-business-update-call -transcript; Ian Urbina, "Drilling Down: Documents: Leaked Industry E-Mails and Reports," accessed September 7, 2011, nytimes.com/interactive /us/natural-gas-drilling-down-documents-4.html, page 74.

12. Chesapeake Energy, "Chesapeake Energy Corporation Announces Sale of Fayetteville Shale Assets to BHP Billiton for $4.75 Billion in Cash," news release, February 21, 2011, accessed July 26, 2011, chk.com/news/articles /pages/1530960.aspx; Chesapeake Energy, "Chesapeake Energy Corporation Provides Operational and Financial Update," news release, September 22, 2008, accessed July 27, 2011, chk.com/news/articles/Pages/1199524 .aspx; Cheseapeake Energy, "February 2011 Investor Presentation," accessed July 27, 2011,slideshare.net/plsderrick/chesapeake-2011-february-investor -presentation, page 19; Chesapeake Energy and BP, "Chesapeake and BP Announce Arkoma Basin Woodford Shale Transaction," news release, July 17, 2008, accessed July 26, 2011 chk.com/news/articles/pages/1176363.aspx.

13. "US Shale and Australian Nickel Asset Review," BHP Billiton Limited, accessed August 28, 2012, bhpbilliton.com/home/investors/news/Pages /Articles/US-Shale-and-Australian-Nickel-Asset-Review.aspx.

14. Reuters, "DAVOS-Shale gas is U.S. energy 'game changer' -BP CEO", January 28, 2010 7:43 am ET, accessed July 26, 2011, reuters.com/article/2010/01/28/davos-energy-idUSLDE60R1MV20100128.

15. Chesapeake Energy and BP, "Chesapeake and BP Announce Arkoma Basin Woodford Shale Transaction," news release, July 17, 2008, accessed July 26, 2011, chk.com/news/articles/pages/1176363.aspx.

16. Ibid.

17. Chesapeake Energy, "Chesapeake Energy Corporation Provides Operational and Financial Update," news release, September 22, 2008, accessed July 27, 2011, chk.com/news/articles/Pages/1199524.aspx.

18. Chesapeake Energy Corporation, "Chesapeake Energy Corporation Reports Financial and Operational Results for the 2012 Second Quarter," news release, August 6, 2012, chk.com/news/articles/Pages/1722883.aspx.

19. Daniel Yergin and Robert Ineson, "America's Natural Gas Revolution," (Wall Street Journal, November 2, 2009), accessed on 6/22/11, online.wsj.com/article/SB10001424052748703399204574507440795971268.html.

20. Mike Soraghan, "Natural Gas: U.S. Fracking Regulation Won't Halt 'Shale Gale'—report" (E&E Publishing LLC, March 10, 2010), accessed on June 22, 2011, eenews.net/public/eenewspm/2010/03/10/1.

21. Daniel Yergin, "Stepping on the Gas," (Wall Street Journal, April 2, 2011), accessed July 2, 2011, online.wsj.com/article/SB10001424052748703712504576232582990089002.html.

22. Daniel Yergin, "It's Not the End Of the Oil Age," (Washington Post, July 31, 2005), accessed July 27, 2011, washingtonpost.com/wp-dyn/content/article/2005/07/29/AR2005072901672.html.

23. bp.com, "BP Statistical Review of World Energy June 2011, Oil Section," accessed June 27, 2011, bp.com/sectiongenericarticle800.do?categoryId=9037157&contentId=7068604, p. 8.

24. Businessweek, "Plenty Of Oil—Just Drill Deeper," September 18, 2006, accessed July 11, 2011, businessweek.com/magazine/content/06_38/b4001055.htm.

25. Association for the Study of Peak Oil & Gas—USA, "Wager Challenges CERA Oil Supply Prediction," February 6, 2008, accessed September 8, 2011, www.peakoil.net/headline-news/aspo-wages-challenges-cesa-oil-supply-prediction.

26. Ibid.

27. Terry Engelder "Marcellus," Fort Worth Oil and Gas Journal, August 2009, accessed September 8, 2011, www3.geosc.psu.edu/~jte2/references/references.html, link 155.

28. John Curtis, PhD, Chairman of the Potential Gas Committee, "Interview with Bill Powers," June 6, 2011, recorded with permission, duration 1:19:46. iTunes Voice Memo format.

29. Pennsylvania Department of Environmental Protection, "Operator Active

Wells," accessed July 5, 2011, dep.state.pa.us/dep/deputate/minres/oilgas /BOGM%20Website%20Pictures/2010/Operator%20Active%20Wells.jpg.

30. US Census, accessed May 12, 2011, quickfacts.census.gov/qfd/states/42/42 015.html.

31. Martin K. Dubois, Alan P. Byrnes, Saibal Bhattacharya, Geoffrey C. Bohling, John H. Doveton, and Robert E. Barba, "Hugoton Asset Management Project (HAMP): Hugoton Geomodel Final Report" (Kansas Geological Survey) (2007).

32. Terry Engelder and Gary G. Lash, "Marcellus Shale Play's Vast Resource Potential Creating Stir In Appalachia," *American Oil & Gas Reporter*, May 2008, aogr.com/index.php/magazine/cover_story_archives/may_2008 _cover_story/.

33. Terry Engelder, "Interview with Bill Powers," July 1, 2011, 1:02 pm, recorded with permission, duration 42:09, iTunes Voice Memo format; Terry Engelder, "Interview with Bill Powers," July 6, 2011 1:09 pm, recorded with permission, duration 13.12, iTunes Voice Memo format.

34. Jonathan D. Silver, "The Marcellus Boom / Origins: the Story of a Professor, a Gas Driller and Wall Street," *Pittsburgh Post-Gazette*, March 20, 2011, accessed September 8, 2011, postgazette.com/pg/11079/1133325-503.stm.

35. Terry Engelder, Professional Vitae, May 14, 2010, accessed May 13, 2011, geosc.psu.edu/academic-faculty/engelder-terry.

36. Timothy Considine, Robert Watson, Rebecca Entler, Jeffrey Sparks, "An Emerging Giant: Prospects and Economic Impacts of Developing the Marcellus Shale Natural Gas Play" (white paper, College of Earth & Mineral Sciences, Department of Energy and Mineral Engineering, Penn State University), July 24, 2009, attached to Jon Bogle, "Penn State Admits Gas Study Flaws," July 1, 2010, accessed August 30, 2012, northcentralpa.com/article /penn-state-admits-gas-study-flaws.

37. Ibid, ii.

38. Ibid., ii.

39. Ibid, ii.

40. Kaustuv Basu, "Fracking Open—Gas Drilling Research Stirs Controversies at Universities," *Inside Higher Ed* (July 6, 2012), accessed August 30, 2012, protecteaglesmere.org/2012/07/07/fracking-open/.

41. Jon Bogle to Dean (William) Easterling, attachment to Jon Bogle, "Penn State Admits Gas Study Flaws," July 1, 2010, accessed August 30, 2012, north centralpa.com/article/penn-state-admits-gas-study-flaws.

42. William Easterling to Drake Saxton, attachment to Jon Bogle, "Penn State Admits Gas Study Flaws," July 1, 2010, accessed August 30, 2012, north centralpa.com/article/penn-state-admits-gas-study-flaws.

43. Ibid.

44. Timothy J. Considine, Robert Watson, Seth Blumsack, "The Economic Impacts of the Pennsylvania Marcellus Shale Natural Gas Play: An Update"

(white paper, College of Earth & Mineral Sciences, Department of Energy and Mineral Engineering, Penn State University), May 24, 2010, accessed August 30, 2012, marcelluscoalition.org/2010/05/the-economic-impacts -of-the-pennsylvania-marcellus-shale-natural-gas-play-an-update; William Easterling to Drake Saxton, attachment to Jon Bogle, "Penn State Admits Gas Study Flaws," July 1, 2010, accessed August 30, 2012, northcentralpa.com /article/penn-state-admits-gas-study-flaws.

45. Timothy Considine, Robert Watson, Rebecca Entler, Jeffrey Sparks, "An Emerging Giant: Prospects and Economic Impacts of Developing the Marcellus Shale Natural Gas Play" (white paper, College of Earth & Mineral Sciences, Department of Energy and Mineral Engineering, Penn State University), July 24, 2009, accessed August 30, 2012, s3.amazonaws.com /propublica/assets/monongahela/EconomicImpactsMarcellus.pdf.

46. Boonepickens.com, "T. Boone Pickens: His Life. His Legacy," accessed July 2, 2011, boonepickens.com/man_ahead/default.asp.

47. *The Business Journal*, "Pickens Backs Renewable Energy Plan," July 8, 2008, accessed July 2, 2011, bizjournals.com/triad/stories/2008/07/07/daily20 .html.

48. Pickensplan.com, "Natural Gas," accessed July 2, 2011, pickensplan.com /natural_gas/.

49. T. Boone Pickens, "T. Boone Pickens Second TV Commercial," , August 1, 2008, accessed September 12, 2011, youtube.com/watch?v=X_3RV5SLS-I& feature=relmfu, duration 0.31, format FLV360P.

50. Govtrack.us, "H.R. 1835: New Alternative Transportation to Give Americans Solutions Act of 2009," accessed July 2, 2011, govtrack.us/congress/bill .xpd?bill=h111-1835&tab=summary.

51. Pickensplan.com, email to author, "Please Sign My Petition Urging President Obama and Congress to Take Action on Energy Reform NOW!," January 8, 2010.

52. T. Boone Pickens, "Interview with T. Boone Pickens on Squawkbox," April 28, 2011, accessed September 12, 2011, youtube.com/watch?v=vsC2SRs7c QE, duration 12:10, FLV480P format.

53. Clean Energy Fuels Corp., "Form 10-K" for period ended December 31, 2011, accessed August 30, 2012, investors.cleanenergyfuels.com/secfiling .cfm?FilingID=1047469-12-2470, page 35.

54. David Lazarus, "Don't Dismiss Pickens' Plan Yet," *Los Angeles Times*, July 9, 2008, accessed August 30, 2012, articles.latimes.com/print/2008/jul/09 /business/fi-lazarus9.

55. Texas Railroad Commission, "Barnett Shale Information—Updated July 20, 2012": Newark, East (Barnett Shale) Statistics, updated 3/05/12, accessed August 31, 2012, rrc.state.tx.us/data/fielddata/barnettshale.pdf; author's estimation.

56. Ibid.
57. Author created estimate using data from Louisiana Department of Natural Resources, accessed September 20, 2011, dnr.louisiana.gov/assets/TAD /data/facts_and_figures/table09.htm; Texas Railroad Commission, accessed September 20, 2011, rrc.state.tx.us/bossierplay/index.php.
58. Ibid.
59. Mark J.Kaiser and Yunke Yu, "Louisiana Haynesville Shale—1: Characteristics, Production Potential of Haynesville Shale Wells Described" (*Oil and Gas Journal*, December 5, 2011), accessed March 6, 2012, ogj.com/articles /print/volume-109/issue-49/exploration-development/louisiana-haynes ville-shale-p1.html, p. 12.
60. Arthur E. Berman and Lynn F. Pittinger, "U.S. Shale Gas: Less Abundance, Higher Cost," accessed, March 9, 2012, theoildrum.com/node/8212.
61. Mark J.Kaiser and Yunke Yu, p. 12.
62. US Energy Information Administration, "Shale Gas Production (Billion Cubic Feet)," accessed June 27, 2011, eia.gov/dnav/ng/ng_prod_shalegas _s1_a.htm; Arkansas Oil and Gas Commission.
63. Author estimate using Arkansas Oil and Gas Commission data, cited in Bill Powers, "A Brief History of a Shale Play," (Powers Energy Investor, September 1, 2011).
64. Author estimate based on data from the Pennsylvania Department of Environmental Protection, accessed May 12, 2011 and September 20, 2011, paoilandgasreporting.state.pa.us/publicreports/Modules/Production /ProductionByCounty.aspx.
65. Texas Railroad Commission, "Eagle Ford Statistics Gas Production Statistics," Updated 12/18/2012, accessed December 22, 2012, rrc.state.tx.us/eagle ford/EagleFordGWGProduction.pdf.
66. Michigan Public Service Commission, "Michigan Natural Gas Production," accessed September 20, 2011, www.dleg.state.mi.us/mpsc/gas/pesec2 .htm.
67. Ibid.
68. Ibid; US Energy Information Administration, "U.S. Natural Gas Wellhead Price (Dollars per Thousand Cubic Feet)," accessed June 7, 2011, eia.gov /dnav/ng/hist/n9190us3a.html.
69. H. G. Dube et al. "SPE 63091: Lewis Shale, San Juan Basin: What We Know Now" (Society of Petroleum Engineers Inc., 2000).
70. Author estimated from reviewing financial statements of producers active in the play.

Chapter 6

1. US Energy Information Administration, Annual Energy Review "Table 8.4a: Consumption for Electricity Generation by Energy Source: Total

(All Sectors), Selected Years, 1949–2009," accessed October 17, 2011, eia.gov/totalenergy/data/annual/txt/ptbo605.html.

2. Ibid; US Energy Information Administration, "Natural Gas Consumption by End Use, Million Cubic Feet," updated 8/31/2012, accessed September 3, 2012, eia.gov/dnav/ng/ng_cons_sum_dcu_nus_a.htm.

3. Ibid.

4. US Energy Information Administration, Annual Energy Review, "Table 9.1: Nuclear Generating Units, 1955–2011," accessed October 17, 2011, eia.gov/totalenergy/data/annual/txt/ptbo901.html; US Energy Information Administration, Annual Energy Review, "Table 7.3: Coal Consumption by Sector, Selected Years 1949–2011," accessed October 17, 2011, eia.gov/totalenergy/data/annual/txt/ptbo703.html.

5. US Energy Information Administration, "Table 8.2a: Electricity Net Generation: Total (All Sectors), Selected Years, 1949–2011," accessed October 17, 2011, eia.gov/totalenergy/data/annual/txt/ptbo802a.html.

6. Ibid.

7. US Energy Information Administration, Annual Energy Review, "Table 8.4a: Consumption for Electricity Generation by Energy Source: Total (All Sectors), Selected Years, 1949–2009," accessed October 17, 2011, www.eia.gov/totalenergy/data/annual/txt/ptbo605.html.

8. Erik Shuster, National Energy Technology Laboratory, "Tracking New Coal-Fired Power Plants," January 14, 2011, accessed October 17, 2011, netl.doe.gov/coal/refshelf/ncp.pdf, slide 7.

9. US Energy Information Administration, Annual Energy Review "Table 9.1: Nuclear Generating Units, 1955–2011," accessed October 17, 2001, eia.gov/totalenergy/data/annual/txt/ptbo901.html.

10. United States Nuclear Regulatory Commission, "Power Uprates for Nuclear Plants," accessed October 17, 2011, nrc.gov/reading-rm/doc-collections/fact-sheets/power-uprates.html.

11. Julie Johnsson, "Nuclear Repairs No Easy Sale as Cheap Gas Hits Utilities," *Bloomberg.com*, September 11, 2012, bloomberg.com/news/2012-09-10/nuclear-repairs-no-easy-sale-as-cheap-gas-hits-utilities.html.

12. Matthew L. Wald, "Wisconsin Nuclear Reactor to Be Closed," *The New York Times*, October 22, 2012, nytimes.com/2012/10/23/business/energy-environment/dominion-to-close-wisconsin-nuclear-plant.html.

13. US Energy Information Administration, "Table 1.1: Net Generation by Energy Source, Thousand Megawatt Hours: Total (All Sectors), Released Date, July 26, 2012, Data from Electric Power Monthly," accessed September 3, 2012, eia.gov/electricity/monthly/xls/table_1_01.xlsx.

14. US Energy Information Administration, "Table 1.2: Existing Capacity by Energy Source, 2009 (Megawatts): Report," revised April 2011, accessed October 17, 2011, eia.gov/electricity/capacity/.

Chapter 7

1. US Energy Information Administration, "U.S. Natural Gas Industrial Consumption (Million Cubic Feet)," accessed October 17, 2011, eia.gov/dnav /ng/hist/n3035us2a.htm; US Energy Information Administration, "U.S. Natural Gas Total Consumption (Million Cubic Feet)," accessed October 17, 2011, eia.gov/dnav/ng/hist/n9140us2a.htm.

2. Wen-yuan Huang, "Factors Contributing to the Recent Increase in US Fertilizer Prices, 2002–08 (United States Department of Agriculture, February 2009), accessed October 17, 2011, ers.usda.gov/Publications/AR33/AR33 .pdf, p. 4.

3. Ibid, p. 6.

4. Ibid, p. 9.

5. Orascom Construction Industries, "OCI Beaumont Operation", accessed January 11, 2013, orascomci.com/index.php?id=pandoramethanolllc.

6. Orascom Cons. Inds., " Orascom Cons. Inds. (ORSD) - OCI Fertilizer Group Selects Iowa for New Plant," news release, September 5, 2012, bloom berg.com/article/2012-09-06/aK31H2deWv84.html.

7. Ibid.

8. Jeff Rubin, *Why Your World Is About To Get a Whole Lot Smaller* (New York: Random House, 2009).

9. Andrew Jacobs, "Honda Strikers in China Offered Less Than Demanded," *New York Times,* June 18, 2010, accessed October 19, 2011, nytimes.com/2010 /06/19/business/global/19strike.html?scp=10&sq=honda+foshan+china +strike&st=nyt.

10. Ibid.

11. "China's Labor Tests Its Muscle," *New York Times,* updated August 16, 2010, accessed October 19, 2011, topics.nytimes.com/top/news/international /countriesandterritories/china/labor-issues/index.html?scp=3&sq=mini mum%20wage%20hong%20kong&st=cse.

12. Don Lee, "Battery Recharges Debate About U.S. Manufacturing," *Chicago Tribute,* May 16, 2010, accessed June 22, 2010, articles.chicagotribune.com /2010-05-16/business/ct-biz-0516-green-manufacture--20100516-5_1_bat tery-lithium-ion-cordless-power-tools.

13. Ibid.

Chapter 8

1. US Energy Information Administration, "Natural Gas Consumption by End Use, release date 8/31/12," accessed September 3, 2012, eia.gov/dnav/ng /ng_cons_sum_dcu_nus_a.htm.

2. US Department of Energy, "Energy Efficiency and Renewable Energy— Bringing You a Prosperous Future Where Energy is Clean, Abundant, Reliable and Affordable," October 2008, accessed October 21, 2011,

apps1.eere.energy.gov/buildings/publications/pdfs/corporate/bt_state industry.pdf, p. 9.

3. Ibid, p. 5.

4. Ibid.

5. US Energy Information Administration, "Natural Gas Deliveries to Commercial Consumers (Including Vehicle Fuel through 1996) in the US (Million Cubic Feet) release date 8/31/2012," accessed September 3, 2012, eia.gov /dnav/ng/hist/n3020us2A.htm; US Energy Information Administration, "Energy Efficiency and Renewable Energy—Bringing You a Prosperous Future Where Energy is Clean, Abundant, Reliable and Affordable," October 2008, accessed October 21, 2011, apps1.eere.energy.gov/buildings/publica tions/.../bt_stateindustry.pdf, p. 6.

6. US Energy Information Administration, "U.S. Natural Gas Residential Consumption (Million Cubic Feet), release date 8/31/12" accessed September 3, 2012, eia.gov/dnav/ng/hist/n3010us2A.htm; US Energy Information Administration, "Natural Gas Deliveries to Commercial Consumers (Including Vehicle Fuel through 1996) in the US (Million Cubic Feet), release date 8/31/12," accessed September 3, 2012, eia.gov/dnav/ng/hist /n3020us2A.htm.

Chapter 9

1. Natural Gas Vehicles for America, "Energy Policy Act of 1992—Fleet Program," accessed October 21, 2011, ngvc.org/gov_policy/fed_regs/fed_EPA FleetPrg.html.

2. Ibid.

3. International Association for Natural Gas Vehicles, "Natural Gas Vehicle Statistics—NGV Count Ranked Numerically As at December 2010," accessed October 21, 2011, iangv.org/current-ngv-stats/.

4. Ibid; US Department of Transportation Research and Innovative Technology Administration, "National Transportation Statistics, 2010," accessed September 3, 2012, bts.gov/publications/national_transportation_statistics /html/table_01_11.html.

5. US Energy Information Administration, "Annual U.S. Natural Gas Vehicle Fuel Consumption (Million Cubic Feet), release date 8/31/2012," accessed September 3, 2012, eia.gov/dnav/ng/hist/n3025us2A.htm; Natural Gas Vehicle Institute, "NGV Connection Newsletter—August 2011," accessed November 30, 2011, ngvi.com/newsletter_august2011.html.

6. Natural Gas Vehicle Institute, "NGV Connection Newsletter—August 2011," accessed November 30, 2011, ngvi.com/newsletter_august2011.html.

7. US Energy Information Administration, "Guidance for Federal Agencies: New Alternative Fuel Vehicle Definitions under Section 2862 of the National Defense Authorization Act of 2008," September 2008, accessed September 3, 2012, www1.eere.energy.gov/femp/pdfs/ndaa_guidance.pdf.

8. US Energy Information Administration, "Annual U.S. Natural Gas Vehicle Fuel Consumption (Million Cubic Feet), release date 8/31/2012," accessed September 3, 2012, eia.gov/dnav/ng/hist/n3025us2A.htm.

9. Jeff Greene, "The Greene Page, The Natural Gas Revival," April 5, 2011, accessed November 30, 2011, wisegasinc.com/wg-greene.htm.

10. Natural NGV Fleet Summit, "Nat Gas Act of 2009—New Alternative Transportation to Give Americans Solutions Act of 2009," accessed November 30, 2011, ngvsummit.com/natgasact.html.

11. Natural NGV Fleet Summit, "Nat Gas Act of 2009—New Alternative Transportation to Give Americans Solitions Act of 2009," accessed November 30, 2011, ngvsummit.com/natgasact.html.

12. US Department of Energy, "Alternative Fuels Data center," accessed January 11, 2013, afdc.energy.gov/fuels/electricity_locations.html.

13. Bill Powers conversation with Luis Giron, Siemens USA Division, August 22, 2012.

14. Solarprices.org, "Benefits of Solar Carports: Green Living in the 21st Century," accessed September 3, 2012, solarprices.org/benefits-of-solar-carports-green-living-in-the-21st-century/#more-115.

15. The Oil Drum blog, "Preditions for Canada's Natural Gas Production," blog entry by "benk," June 4, 2008 at 10:00 am, accessed November 30, 2011, theoildrum.com/node/4073.

16. Kenneth S. Deffeyes, *Beyond Oil—The View from Hubbert's Peak* (New York: Hill and Wang, a division of Farrar, Straus and Giroux, 2005), p. 79.

Part 3

1. US Energy Information Administration, "U.S. Natural Gas Marketed Production (Million Cubic Feet), 1900–2011," accessed March 15, 2012, eia.gov/dnav/ng/hist/n9050us2a.htm.

2. Ibid.

3. US Energy Information Administration, "Natural Gas Wellhead Value and Marketed Production," released September 28, 2012, pulled for states: Texas: eia.gov/dnav/ng/ng_prod_whv_dcu_STX_a.htm; Louisiana: eia.gov/dnav/ng/ng_prod_whv_dcu_SLA_a.htm; Federal Offshore/ Gulf of Mexico: eia.gov/dnav/ng/hist/n9050fx2a.htm; Wyoming: eia.gov /dnav/ng/ng_prod_whv_dcu_SWY_a.htm; New Mexico: eia.gov/dnav /ng/ng_prod_whv_dcu_SNM_a.htm; Arkansas: eia.gov/dnav/ng/ng_ prod_whv_dcu_SAR_a.htm; Pennsylvania: Pennsylvania Department of Environmental Protection, "Statewide Data Downloads by Reporting Period—Jan–Jun 2011 (Marcellus Only, 6 months)," accessed September 8, 2012, paoilandgasreporting.state.pa.us/publicreports/Modules/Data Exports/ExportProductionData.aspx?PERIOD_ID=2011-1; Pennsylvania Department of Environmental Protection, "Statewide Data Downloads by Reporting Period—Jul–Dec 2011 (Marcellus Only, 6 months)," accessed

September 8, 2012, paoilandgasreporting.state.pa.us/publicreports/Mod
ules/DataExports/ExportProductionData.aspx?PERIOD_ID=2011-2.

Chapter 10

1. US Energy Information Administration, "Natural Gas Gross Withdrawals
   and Production, Area: U.S.—2011, release Date: 8/31/2012," accessed September 3, 2012, eia.gov/dnav/ng/ng_prod_sum_dcu_NUS_a.htm; US
   Energy Information Administration, "Natural Gas Gross Withdrawals and
   Production, Area: Texas—2011, release Date 8/31/2012," accessed September 3, 2012, eia.gov/dnav/ng/ng_prod_sum_dcu_stx_a.htm.
2. bp.com, "BP Statistical Review of World Energy June 2012," accessed September 3, 2012, bp.com/sectionbodycopy.do?categoryId=7500&contentId
   =7068481; bp.com/assets/bp_internet/globalbp/globalbp_uk_english
   /reports_and_publications/statistical_energy_review_2011/STAGING
   /local_assets/pdf/statistical_review_of_world_energy_full_report_2012
   .pdf, p. 22; Texas Railroad Commission, "Texas Monthly Oil and Gas Production (January 2006–May 2012) updated 7/25/12," accessed September
   4, 2012, rrc.state.tx.us/data/production/index.php, rrc.state.tx.us/data
   /production/ogismcon.pdf.
3. Texas Railroad Commission, "Oil and Gas Division District Boundaries,"
   accessed October 21, 2011, rrc.state.tx.us/forms/maps/districts_colorsm
   .jpg.
4. David Michael Cohen, "Eagle Ford: Texas' Dark-Horse Resource Play Picks
   up Speed," *World Oil Online*, Vol. 232, No. 6, June 2011, accessed September
   4, 2012, worldoil.com/June-2011-Eagle-Ford-Texas-dark-horse-resource
   -play-picks-up-speed.html.
5. Texas Railroad Commission, "Eagle Ford Information," updated 08/30/12,
   accessed August 31, 2012, rrc.state.tx.us/eagleford/index.php.
6. US Energy Information Administration, "Eagle Ford Shale Play, Western
   Gulf Basin, South Texas," map date: May 29, 2010, accessed October 24,
   2011, eia.gov/oil_gas/rpd/shaleusa9.pdf.
7. "Two South Texas Pipelines to Handle Eagle Ford Shale Production," *Pipeline & Gas Journal*, February 2011, Vol. 238, No. 2, accessed September 4,
   2012, pipelineandgasjournal.com/two-south-texas-pipelines-handle-eagle
   -ford-shale-production; PRNewswire, "Kinder Morgan-Copano Increase
   Presence in Eagle Ford Shale With New Long Term Contracts," June 30,
   2011, accessed September 4, 2012, redorbit.com/news/business/2072981
   /kinder_morgancopano_increase_presence_in_eagle_ford_shale_with
   _new/.
8. Texas Railroad Commission, "Eagle Ford Fields and Counties 12-5-2012,"
   accessed December 23, 2012, rrc.state.tx.us/eagleford/EagleFord_Fields
   _and_Counties_201212.xls.
9. Texas Railroad Commission, "Natural Gas Production and Well Counts

(1935–2011) History of Texas Intial Crude Oil, Annual Production and Producing Wells," accessed September 9, 2012, rrc.state.tx.us/data/production/gaswellcounts.php.

10. Texas Railroad Commission, "General Production Query Results," District 1 Annual Data 2008–2011, accessed September 20, 2012, webapps.rrc .state.tx.us/PDQ/changeViewReportAction.do?viewType=Annual%20 Totals.

11. Texas Railroad Commission, "General Production Query Results," Districts 2, 3 and 4, Annual Data 2008–2011, accessed September 12, 2012, webapps .rrc.state.tx.us/PDQ/changeViewReportAction.do?viewType=Annual%20 Totals.

12. Sandra Sue Sodersten and Debasish Sihi, Shell E&P Co.; and Marilyn Taggi Cisar, Shell EP-Americas, "HPHT, Low Permeability Fields in South Texas.... A Legacy of Development in a Tough Environment." Abstract. *International Petroleum Technology* Conference (Dubai, U.A.E., December 4–6, 2007), accessed March 15, 2012, onepetro.org/mslib/servlet/onepetrop review?id=IPTC-11581-ABSTRACT&soc=IPTC.

13. Ibid.

14. Hearing on the application of Mitchell Energy Company for Exception to Statewide Rule 23(a)(2) for Various Leases in Boonesville (Bend Congl., Gas) Field, Wise County, Texas, September 13, 2001, Oil and Gas Docket No. 09-0228612, before Margaret Allen, Technical Hearings Examiner, rrc.state.tx.us/meetings/ogpfd/documents/09-28612pfd.pdf.

15. Texas Railroad Commission, "General Production Query Results," District 5, Annual Data 2008–2011, accessed September 12, 2012, webapps.rrc .state.tx.us/PDQ/changeViewReportAction.do?viewType=Annual%20 Totals.

16. Texas Railroad Commission, "General Production Query Results," District 6, Annual Data 2008–2011, accessed September 12, 2012, webapps.rrc .state.tx.us/PDQ/changeViewReportAction.do?viewType=Annual%20 Totals.

17. Baker Hughes, "North America Rotary Rig Counts Current Week Data—Rigs by State—Current & Historical," accessed September 7, 2012, investor .shareholder.com/common/download/download.cfm?companyid=BHI &fileid=598392&filekey=9CD5C2B7-07B1-451E-8385-B40D82FBD5BB& filename=Rigs_by_State_090712.xlsx.

18. Texas Railroad Commission, "General Production Query Results," District 8, Monthly Totals 2008–2011, accessed September 12, 2012, webapps.rrc.state .tx.us/PDQ/changeViewReportAction.do?viewType=Monthly%20Totals.

19. Texas Railroad Commission, "General Production Query Results," District 8, Annual Data 2008–2011, accessed September 12, 2012, webapps.rrc.state.tx .us/PDQ/changeViewReportAction.do?viewType=Annual%20Totals.

20. Ibid.

21. Texas Railroad Commission, "Barnett Shale Information, Statistics: Newark, East (Barnett Shale) Statistics, updated 3/5/2012," accessed September 12, 2012, rrc.state.tx.us/data/fielddata/barnettshale.pdf.

22. Texas Railroad Commission, "Barnett Shale Information, Texas Gas Well Gas Production in Newark East (Barnett Shale) Field—1993–2011, updated 7/20/12," accessed September 12, 2012, rrc.state.tx.us/barnettshale/Newark EastField_1993-2011.pdf.

23. Texas Railroad Commission, "Barnett Shale Information, Statistics: Newark, East (Barnett Shale) Statistics, updated 3/5/2012," accessed September 12, 2012, rrc.state.tx.us/data/fielddata/barnettshale.pdf.

24. Texas Railroad Commission, "General Production Query Results," District 9, Annual Data 2006–2011, accessed September 12, 2012, webapps.rrc.state.tx .us/PDQ/changeViewReportAction.do?viewType=Annual%20Totals.

25. Texas Railroad Commission, "Barnett Shale Information, Texas Counties with Producing Wells and Contact Information," updated 2/17/2012, accessed September 12, 2012, rrc.state.tx.us/barnettshale/images/county producing2-sm.php.jpg.

26. Texas Railroad Commission, "Barnett Shale Information, Statistics: Newark East (Barnett Shale) Statistics," updated 3/5/2012, accessed September 12, 2012, rrc.state.tx.us/data/fielddata/barnettshale.pdf.

27. GlobalNewswire via Comtex News Network, "LINN Energy Announces 60.2 MMcfe Per Day Horizontal Granite Wash Well," news release, July 22, 2010, ir.linnenergy.com/releasedetail.cfm?ReleaseID=491038.

28. Business Wire, "Chesapeake Energy Corporation Announces Significant New Discovery in the Hogshooter Play of the Texas Panhandle and Western Oklahoma," news release, June 1, 2012, chk.com/News/Articles/Pages /1701619.aspx.

29. Texas Railroad Commission, "General Production Query Results," District 10, Annual Data 2008–2011, accessed September 12, 2012, webapps.rrc .state.tx.us/PDQ/changeViewReportAction.do?viewType=Annual%20 Totals.

30. Texas Railroad Commission, "Natural Gas Production and Well Counts (1935–2011)," accessed September 12, 2012, rrc.state.tx.us/data/production /gaswellcounts.php.

31. Texas Railroad Commission, "Monthly Oil and Gas Production by Year, 2006–2012, updated 7/25/2012," accessed September 12, 2012, rrc.state.tx.us /data/production/ogismcon.pdf.

32. Texas Railroad Commission, "Natural Gas Production and Well Counts (1935–2011)," accessed September 12, 2012, rrc.state.tx.us/data/production /gaswellcounts.php.

33. Ibid.

Chapter 11

1. State of Louisiana, Department of Natural Resources, "Louisiana State Gas Production, Wet After Lease Separation," revised June 4, 2012, accessed September 17, 2012, dnr.louisiana.gov/assets/TAD/data/facts_and_figures /table12.htm.
2. Ibid.
3. Ibid.
4. History of Caddo Parish, "Oils and Natural Gas," accessed September 21, 2012, caddohistory.com/oil_gas.html.
5. Louisiana Mid-Continent Oil and Gas Association, "History of the Industry—Oil and Gas 101," accessed September 21, 2012, lmoga.com/resources /oil-gas-101/history-of-the-industry.
6. State of Louisiana, Department of Natural Resources, "Louisiana State Gas Production, Wet After Lease Separation," revised June 4, 2012, accessed September 17, 2012, dnr.louisiana.gov/assets/TAD/data/facts_and_figures /table12.htm.
7. Ibid.
8. Louisiana Bureau of Land Management, "Reasonably Foreseeable Development Scenario for Fluid Minerals," March 2008, blm.gov/pgdata/etc /medialib/blm/es/jackson_field_office/planning/planning_PDF_as_rfds .Par96360.File.dat/LA_RFDS_R2.pdf.
9. Ibid, p. 4.
10. Peggy Williams, "Prolific and Proud: A Century of E&P in Louisiana," *Oil and Gas Investor*, A supplement, 2005, oilandgasinvestor.com/pdf/LASuple ment.pdf, p. 11.
11. The Paleontological Research Institute, "Salt Dome Trap," accessed September 23, 2012, priweb.org/ed/pgws/systems/traps/structural/structural .html#salt.
12. James Moffett, McMoRan Exploration, The Oil & Gas Conference Presentation, August 15, 2011.
13. Don Stowers, "Davy Jones discovery may herald new wave of drilling on GoM shelf," *Oil & Gas Financial Journal*, March 1, 2010, ogfj.com/index /article-display.articles.oil-gas-financial-journal.e-__p.Davy-Jones-discovery -may-herald-new-wave-of-drilling-on-GoM-shelf.QP129867.dcmp=rss.page =1.html.
14. David Joint, McMorRan Exploration, conversation with author, September 17, 2012.
15. Mark Kaise, Yunke Yu, "LOUISIANA HAYNESVILLE SHALE—1: Characteristics, production potential of Haynesville shale wells described," December 5, 2011, ogj.com/articles/print/volume-109/issue-49/exploration -development/louisiana-haynesville-shale-p1.html.

16. Bob Tippee, "Shale Gas supply expected to keep US prices low in 2011," *Oil and Gas Journal*, December 3, 2010, pennenergy.com/index/articles /pe-article-tools-template/_saveArticle/articles/oil-gas-journal/drilling -production-2/2010/12/shale-gas_supply_expected.html.

17. Starr Spencer, "Haynesville Shale Primed to Become World's Largest Gas Field by 2020," *Platts* republished on rigzone.com, February 11, 2009, rigzone.com/news/article.asp?a_id=72839.

18. Mark Kaise, Yunke Yu, "Louisiana Haynesville Shale—1: Characteristics, Production Potential of Haynesville Shale Wells Described," December 5, 2011, ogj.com/articles/print/volume-109/issue-49/exploration-develop ment/louisiana-haynesville-shale-p1.html.

19. Ibid.

20. Louisiana Department of Natural Resources, "Haynesville Shale Wells Ac-tivity by Month," September 4, 2012, dnr.louisiana.gov/assets/OC/haynes ville_shale/haynesville_monthly.pdf.

21. Louisiana Department of Natural Resources, "Haynesville Shale Gas Play Well Activity Map," updated September 20, 2012, accessed September 26, 2012, dnr.louisiana.gov/assets/OC/haynesville_shale/haynesville_20111 216.jpg.

22. State of Louisiana, Department of Natural Resources, "Louisiana State Gas Production, Wet After Lease Separation," revised June 4, 2012, accessed September 17, 2012, dnr.louisiana.gov/assets/TAD/data/facts_and_figures /table12.htm.

Chapter 12

1. US Energy Information Administration, "Federal Offshore Gulf of Mexico, Natural Gas Gross Withdrawals and Production," released September 28, 2012, accessed September 28, 2012, eia.gov/dnav/ng/ng_prod_sum_dcu _r3fm_m.htm.

2. Ibid, p. v.

3. Ibid.

4. Ibid.

5. US Department of the Interior, Minerals Management Service, Gulf of Mexico OCS Region, "Outer Continental Shelf Estimated Oil and Gas Reserves, Gulf of Mexico," December 31, 2006, OCS Report MMS 2009-064, boem.gov/uploadedFiles/BOEM/Oil_and_Gas_Energy_Program /Resource_Evaluation/Reserves_Inventory/2009-064GOMR%281%29 .pdf, p. 45.

6. Ibid, p. 33.

7. Baker Hughes, "North America Rotary Rig Counts Through 2011," June 22, 2012, accessed September 29, 2012, investor.shareholder.com/common /download/download.cfm?companyid=BHI&fileid=579035&filekey=A3

A9520E-DDE5-4682-ADBC-B384AC067849&filename=North_America _Rotary_Rig_Counts_through_2011.xls; US Energy Information Administration, "Historical Natural Gas Annual 1930 through 2000," (1992– 1996), December 2001, eia.gov/oil_gas/natural_gas/data_publications /historical_natural_gas_annual/hnga.html; US Energy Information Administration, Federal Offshore—Gulf of Mexico Natural Gas Marketed Production," (1997-2011), released September 28, 2012, accessed September 29, 2012, eia.gov/dnav/ng/hist/n9050fx2a.htm; US Energy Information Administration, "U.S. Natural Gas Wellhead Price," released September 28, 2012, accessed September 29, 2012, eia.gov/dnav/ng/hist /n9190us3a.htm.

8. US Energy Information Administration, "Historical Natural Gas Annual 1930 through 2000," (1992–1996), December 2001, eia.gov/pub/oil_gas /natural_gas/data_publications/historical_natural_gas_annual/current /pdf/hnga00.pdf; US Energy Information Administration, Federal Offshore—Gulf of Mexico Natural Gas Marketed Production," (1997–2011), released September 28, 2012, accessed September 29, 2012, eia.gov/dnav/ng /hist/n9050fx2a.htm.

Chapter 13

1. US Energy Information Administration, "Wyoming Natural Gas Marketed Production—Annual data," released August 31, 2012, accessed September 26, 2012, eia.gov/dnav/ng/hist/n9050wy2a.htm.
2. Ibid.
3. Wyoming Oil and Gas Conservation Commission, Production query system, "Report Date: 9/29/2012, Production for Year 2011 Based on Gas Production—Top 30 producing fields reviewed," wogcc.state.wy.us; Wyoming Oil and Gas Conservation Commission, Production query system, "Report Date: 9/29/2012, Production for Year 2011 Based on Gas Production—All producing fields reviewed," wogcc.state.wy.us.
4. Ultra Petroleum, "Location," accessed September 26, 2012, ultrapetroleum .com/Our-Properties/Green-River-Basin/Location-17.html.
5. "Producing Pinedale with a minimal footprint," *Petroleum Economist*, August 1, 2008, accessed on business.highbeam.com, December 20, 2012, business.highbeam.com/435608/article-1G1-184291658/producing-pinedale -minimal-footprint.
6. Rena Delbridge, "No Guts, No Glory: John Martin's Hunches Have Paid Off in a Big Way for Wyoming," *Made in Wyoming*, accessed September 26, 2012, madeinwyoming.net/profiles/martin.php.
7. "Encana Proposes Up to 3,500 New Wells Near Jonah Field," Wyoming Energy News, April 15, 2011, wyomingenergynews.com/2011/04/encana -proposes-up-to-3500-new-wells-near-jonah-field/; Wyoming Oil and

Gas Conservation Commission, Production query system, "Report Date: 9/30/2012, Production for Year 1997 and 2011 Based on Gas Production—All fields reviewed," wogcc.state.wy.us.

8. Wyoming Oil and Gas Conservation Commission, Production query system, "Report Date: 9/30/2012, Production for Year 2011 Based on Gas Production—All fields reviewed," wogcc.state.wy.us.

9. Pinedale Online, "Pinedale—Jonah Locator Map, Southwest Wyoming," accessed September 26, 2012, pinedaleonline.com/news/2011/05/GasFields.htm.

10. Wyoming Oil and Gas Conservation Commission, Production query system, "Report Date: 9/30/2012, Production for Year 1992 and 2011 Based on Gas Production—All producing fields reviewed," wogcc.state.wy.us.

11. Ultra Petroleum, "About Us," accessed September 29, 2012, ultrapetroleum.com/About-Us-4.html; Ultra Petroleum, "Geologic Setting," accessed September 29, 2012, ultrapetroleum.com/Our-Properties/Green-River-Basin/Geologic-Setting-18.html.

12. Wyoming Oil and Gas Conservation Commission, Production query system, "Report Date: 9/30/2012, Production for Year 1992 and 2011 Based on Gas Production—All producing fields reviewed," wogcc.state.wy.us.

13. "Encana Proposes Up to 3,500 New Wells Near Jonah Field," Wyoming Energy News, April 15, 2011, wyomingenergynews.com/2011/04/encana-proposes-up-to-3500-new-wells-near-jonah-field/.

14. US Geological Survey, map graphic used in Dustin Bleizeffer, "CBM: Bugs vs. Bankruptcy," WyoFile.com, accessed September 28, 2012, wyofile.com/2011/12/california-company-wyo-legislator-seek-to-delay-well-plugging-with-hopes-that-microbes-will-rebuild-production/.

15. Thomas E. Doll, "Coalbed Methane 2009 Update Powder River Basin," Wyoming Oil and Gas Conservation Commission, July 6, 2009, deq.state.wy.us/out/downloads/Eggerman%20prbwater.pdf, p. 5.

16. Wyoming Oil and Gas Conservation Commission, Production query system, "Report Date: 9/30/2012, Production for Year 1997 and 2011 Based on Gas Production—All fields reviewed," wogcc.state.wy.us.

17. Wyoming Oil and Gas Conservation Commission, "Wyoming CBM Production MCF," accessed September 28, 2012, wogcc.state.wy.us/StateCbmGraph.cfm.

18. Wyoming Oil and Gas Conservation Commission, Production query system, "Report Date: 9/30/2012, Production for Year 1992 and 2011 Based on Gas Production—All producing fields reviewed," wogcc.state.wy.us.

19. *North American Unconventional Gas Market Report 2007* (Warlick International, 2007), p. 163; *Management and Effects of Coalbed Methane Produced Water in the United States*, Committee on Management and Effects of Coalbed Methane Development and Produced Water in the Western United

States, A Committee on Earth Resources, National Research Council (2010, Washington, DC: The National Academies Press), nap.edu/openbook.php ?record_id= 12915, p. 93.

20. *Management and Effects of Coalbed Methane Produced Water in the United States*, Committee on Management and Effects of Coalbed Methane Development and Produced Water in the Western United States, A Committee on Earth Resources, National Research Council (2010, Washington, DC: The National Academies Press), nap.edu/openbook.php?record_id =12915, p. 93.

21. Ibid.

22. J.K. Hunter, "Abstract of: LaBarge Project: Availability of CO2 for Tertiary Projects," *Journal of Petroleum Technology*, Vol. 39, No. 11, (November 1987) onepetro.org/mslib/servlet/onepetropreview?id=00015160&soc=SPE; Rod De Bruin, email message to author, February 7, 2011.

23. Rod De Bruin, email message to author, February 18, 2011; State Board of Equalization for the State of Wyoming v. ExxonMobil Corporation, Tax Appeal, Docket No. 2006-69 and Docket No. 2006-116, taxappeals.state.wy .us/images/docket_no_200669etal.htm.

24. Rod De Bruin, email message to author, February 7, 2011.

25. Rod De Bruin, email message to author, February 7, 2011.

26. Peggy Williams, "Wyoming—Operators in the Equality State are Driving Gas Production to New Highs, Thanks to Vigorous Drilling Programs in Both Old and New Areas," *Oil and Gas Investor*, March 2000, unitcorp.com /pdf/Wyoming.pdf, p. 32.

27. Ibid.

28. Ibid.

29. Dustin Bleizeffer, "Gas Plant Still Idle After Fire," *Casper Star-Tribune*, June 9, 2010, trib.com/news/state-and-regional/article_ee198bda-27f7-54e4-a9 96-22dfab1d54ea.html.

30. Evelyn Pyburn, "Denbury to Invest Billions in CO_2 Oil Recovery in Montana," *Big Sky Business Journal*, April 18, 2012, bigskybusiness.com/index.php /business/economy/2511-denbury-to-invest-billions-in-co2-oil-recovery-in -montana.

31. US Energy Information Administration, "Figure 13.05 Wyoming Natural Gas Marketed Production - Annual," release date 9/28/2012, accessed September 30, 2012, eia.gov/dnav/ng/hist/n9050wy2a.htm.

Chapter 14

1. US Energy Information Administration, "New Mexico Natural Gas Wellhead Value and Marketed Production," release date 8/31/2012, accessed September 6, 2012, eia.gov/dnav/ng/xls/NG_PROD_WHV_DCU_SNM _A.xls.

2. New Mexico Bureau of Geology and Mineral Resources, "Coalbed Methane in New Mexico," *New Mexico Earth Matters*, Vol 4, No. 1, 2004, geoinfo .nmt.edu/publications/periodicals/earthmatters/4/EMV4N1.pdf.

3. Kathy Shirley, "Potential is Now a Reality—Coalbed Methane Comes of Age," *Explorer*, March 2000, accessed September 6, 2012, aapg.org/explorer /2000/03mar/coalbedmeth.cfm.

4. Alex Chakhmakhchev, Bob Fryklund, "Critical Success Factors of CBM Development—Implications of two strategies to global development," 19th World Petroleum Congress, Spain, 2008, onepetro.org/mslib/servlet/one petropreview?id=WPC-19-3486.

5. Ibid.

6. James E. Fassett, "The San Juan Basin, a Complex Giant Gas Field, New Mexico and Colorado," *Search and Discovery*, #10254, August 31, 2010, search anddiscovery.com/documents/2010/10254fassett/ndx_fassett.pdf, p. 33.

7. Ibid; US Energy Information Administration, "US Coalbed Methane— Past, Present and Future," November 2007, accessed September 6, 2012, eia.gov/oil_gas/rpd/cbmusa2.pdf.

8. Alex Chakhmakhchev, Bob Fryklund, "Critical success factors of CBM development—Implications of two strategies to global development," 19th World Petroleum Congress, Spain 2008, onepetro.org/mslib/servlet/one petropreview?id=WPC-19-3486.

9. New Mexico Oil Conservation Division, "Natural Gas and Oil Production [Friday, August 31, 2012]," accessed September 6, 2012, https://wwwapps .emnrd.state.nm.us/ocd/ocdpermitting/Reporting/Production/Produc tionInjectionSummaryReport.aspx.

10. Ibid.

11. James E. Fassett, "The San Juan Basin, a Complex Giant Gas Field, New Mexico and Colorado," *Search and Discovery*, #10254, August 31, 2010, search anddiscovery.com/documents/2010/10254fassett/ndx_fassett.pdf, p. 1.

12. Ibid, p. 9.

13. R. M. Flores and L. R. Bader, "A Summary of Tertiary Coal Resources of the Raton Basin, Colorado and New Mexico," US Geological Survey Professional Paper 1625-A, pubs.usgs.gov/pp/p1625a/Chapters/SR.pdf.

14. New Mexico Oil Conservation Division, "Natural Gas and Oil Production [Friday, August 31, 2012]," accessed September 6, 2012, https://wwwapps .emnrd.state.nm.us/ocd/ocdpermitting/Reporting/Production/Produc tionInjectionSummaryReport.aspx.

15. Ibid.

Chapter 15

1. Frank Wicks, "The Oil Age," *Mechanical Engineering, the Magazine of ASME*, August 2009, memagazine.asme.org/Articles/2009/August/Oil_Age.cfm.

2. John Harper, "The Marcellus Shale—An Old 'New' Gas Reservoir in Pennsylvania," *Pennsylvania Geology*, Vol. 38, No. 1, Spring 2008, dcnr.state.pa.us /cs/groups/public/documents/document/dcnr_006811.pdf, p. 3.

3. Ibid.

4. Ibid.

5. Jaime Kostelnik and Kristin M. Carter, "The Oriskany Sandstone Updip Permeability Pinchout: A Recipe for Gas Production in Northwestern Pennsylvania?," *Pennsylvania Geology*, Vol. 39, No. 4, Winter 2009, dcnr.state .pa.us/cs/groups/public/documents/document/dcnr_006816.pdf, p. 20.

6. Ibid.

7. Ibid, p. 19.

8. John Harper, "The Marcellus Shale—An Old 'New' Gas Reservoir in Pennsylvania," *Pennsylvania Geology*, Vol. 38, No. 1, Spring 2008, dcnr.state.pa.us /cs/groups/public/documents/document/dcnr_006811.pdf, p. 3.

9. Ibid., p. 8.

10. Range Resources Corporation,"Range Pioneered the Marcellus Shale Play in 2004 with the Successful Drilling of a Vertical Well, the Renz #1," accessed September 8, 2012, rangeresources.com/Operations/Marcellus -Division.aspx.

11. John Harper and Jaime Kostelnik, "The Marcellus Shale Play in Pennsylvania," Pennsylvania Geological Survey, accessed September 8, 2012, marcellus .psu.edu/resources/PDFs/DCNR.pdf, p. 38.

12. "Chevron Acquires Marcellus Acreage from Chief Oil and Tug Hill. Cost May be Between \$7,000-\$11,000/acre!," May 5, 2011, accessed September 8, 2012, mergersandacquisitionreviewcom.blogspot.com/2011/05/chevron -acquires-marcellus-acreage-from.html.

13. US Energy Information Administration, "Pennsylvania Natural Gas Marked Production," released August 31, 2012, accessed September 8, 2012, eia.gov /dnav/ng/hist/n9050pa2A.htm; Pennsylvania Department of Environmental Protection, "Statewide Data Downloads by Reporting Period—Jan–Jun 2011 (Marcellus Only, 6 months)," accessed September 8, 2012, paoiland gasreporting.state.pa.us/publicreports/Modules/DataExports/Export ProductionData.aspx?PERIOD_ID=2011-1; Pennsylvania Department of Environmental Protection, "Statewide Data Downloads by Reporting Period—Jul–Dec 2011 (Marcellus Only, 6 months)," accessed September 8, 2012, paoilandgasreporting.state.pa.us/publicreports/Modules/Data Exports/ExportProductionData.aspx?PERIOD_ID=2011-2; Pennsylvania Department of Environmental Protection, "Statewide Data Downloads by Reporting Period—Jan–Dec 2011 (Annual O&G, without Marcellus)," accessed September 8, 2012, paoilandgasreporting.state.pa.us/publicreports /Modules/DataExports/ExportProductionData.aspx?PERIOD_ID=2011 -0; Pennsylvania Department of Environmental Protection, "Statewide

Data Downloads by Reporting Period—Jan–Jun 2012 (Unconventional Wells)," accessed September 8, 2012, paoilandgasreporting.state.pa.us/pub licreports/Modules/DataExports/ExportProductionData.aspx?PERIOD _ID=2012-1.

14. Chesapeake Energy, "Chesapeake Energy Corporation Reports Financial and Operational Results for the 2012 First Quarter," news release, May 1, 2012, accessed September 9, 2012, chk.com/news/articles/Pages/1689968.aspx.

15. Smita Madhur, "Update 2-RLPC: Chesapeake increases bridge loan to $4B vs $3B, *Reuters*, May 15, 2012, accessed September 9, 2012, chk.com/news /articles/Pages/1689968.aspx.

16. "Shooters—a 'Fracking' History," American Oil & Gas Historical Society, accessed September 9, 2012, aoghs.org/technology/shooters-well-fracking -history/.

17. Al Granberg, illustration in "What is Hydraulic Fracturing," ProPublica.org, accessed September 9, 2012, propublica.org/special/hydraulic-fracturing -national.

18. Erika Staaf, "Risky Business: An Analysis of Marcellus Shale Gas Drilling Violations in Pennsylvania 2008–2011," PennEnvironment Research & Policy Center, February 2012, accessed September 9, 2012, pennenvironment center.org/sites/environment/files/reports/Risky%20Business%20Violat ions%20Report_0.pdf.

19. Ibid.

20. Tom Barnes, "2 drillers fined for Pennsylvania gas well blowout," *Pittsburgh Post-Gazette*, March 29, 2012, post-gazette.com/stories/local/state/2-drillers -fined-for-pennsylvania-gas-well-blowout-255250.

21. Department of Environmental Protection, "DEP Fines Chesapeake Appalachia $565,000 for Multiple Violations, news release, February 9, 2012, portal .state.pa.us/portal/server.pt/community/newsroom/14287?id=19258&type id=1.

22. Laura Olson, "Pa. fines driller $1.1 million over contamination, fire," *Pittsburgh Post-Gazette*, March 30, 2012, post-gazette.com/stories/local/break ing/pa-fines-driller-11-million-over-contamination-fire-298109/.

23. PRNewswire-USNewswire, "PA DEP Takes Aggressive Action Against Cabot Oil & Gas Corp. to Enforce Environmental Laws, Protect Public in Susquehanna County," news release, April 15, 2012, prnewswire.com/news -releases/pa-dep-takes-aggressive-action-against-cabot-oil--gas-corp-to -enforce-environmental-laws-protect-public-in-susquehanna-county-909 51864.html.

24. "Methane Levels 17 Times Higher in Water Wells Near Hydrofracking," news release, Duke Nicholas School of the Environment, May 9, 2011, accessed September 9, 2012, nicholas.duke.edu/hydrofracking/methane -levels-17-times-higher-in-water-wells-near-hydrofracking-sites.

25. Ibid.
26. Stephen G. Osborn, Avner Vengosh, Nathaniel R. Warner, and Robert B. Jackson, "Methane Contamination of Drinking Water Accompanying Gas Well Drilling and Hydraulic Fracturing," Center on Global Change, Nicholas School of the Environment, Division of Earth and Ocean Sciences, Nicholas School of the Environment, and Biology Department, Duke University, Durham, NC 27708, January 13, 2011, nicholas.duke.edu/hydro fracking/resolveuid/eb4037ebd7c508e8a371f22bde2d4ac8, p. 2.
27. Ibid.
28. Ibid.
29. Ibid.
30. Robert Jackson and Avner Vengosh, "DEP: Protecting water or gas?," December 2, 2011, articles.philly.com/2011-12-02/news/30467569_1_drinking -water-water-resources-methane.

Chapter 16

1. Robb M. Stewart, "BHP Billiton's Petrohawk Swings to Lose," *The Wall Street Journal—Market Watch*, August 9, 2012, marketwatch.com/story /bhp-billitons-petrohawk-swings-to-loss-2012-08-09; BP, "BP p.l.c Group Results, Second Quarter and Half Year 2012(a)," July 31, 2012, bp.com/live assets/bp_internet/globalbp/STAGING/global_assets/downloads/B/bp _second_quarter_2012_results.pdf, p. 6.
2. Author estimate based on US Energy Information Administration, "Natural Gas Gross Withdrawals and Production," released September 28, 2012, eia .gov/dnav/ng/ng_prod_sum_dcu_sar_a.htm; author estimate based on Geology.com, "Fayetteville Shale Orientation," accessed October 21, 2012, geology.com/articles/fayetteville-shale.shtml.
3. John C. Gargani, "Dynamics of Starting a New Resource Play—Fayetteville Shale," November 18, 2008, plsx.com/finder/viewer.aspx?q=&f=src.100&v =0&doc=6240&slide=127701#doc=6240&slide=127701&.
4. Ibid.
5. Arkansas Oil and Gas Commission, "Fayetteville Shale Information—Well Gas Sales Reports—Lifetime Totals—All Counties 7/31/2012," accessed November 1, 2012, aogc2.state.ar.us/FayettevilleShaleInfo/regularly%20up dated%20docs/B-43%20Field%20-%20Well%20MCF%20Totals.xls.
6. Southwestern Energy, "October 2012 Update," October 2012, swn.com/in vestors/LIP/latestinvestorpresentation.pdf, p. 8.
7. Chesapeake Energy Corporation, "Chesapeake Energy Corporation Provides Quarterly Operational Update," news release, August 2, 2010, chk.com /News/Articles/Pages/1455265.aspx.
8. Arkansas Oil and Gas Commission, "Fayetteville Shale Information—Well Gas Sales Reports—Lifetime Totals—All Counties 7/31/2012," accessed

November 1, 2012, aogc2.state.ar.us/FayettevilleShaleInfo/regularly%20 updated%20docs/B-43%20Field%20-%20Well%20MCF%20Totals.xls.

9. Ibid.

10. Chesapeake Energy Corporation, "Chesapeake Energy Corporation Announces Sale of Fayetteville Shale Assets to BHP Billiton for $4.75 Billion in Cash," news release, February 21, 2011, chk.com/news/articles/pages/1530 960.aspx; *Oil & Gas Journal*, "BHP Billiton Writes Down Fayetteville Shale Value," August 3, 2012, ogj.com/articles/2012/08/bhp-billiton-writes-down -us-shale-values.html.

11. "Southwestern Energy's CEO Discusses Q4 2010, Results—Earnings Call Transcript," posted on SeekingAlpha.com, February 25, 2011, seekingalpha .com/article/255116-southwestern-energy-s-ceo-discusses-q4-2010-results -earnings-call-transcript?part=single.

12. Arkansas Oil and Gas Commission, "Fayetteville Shale Information—Well Gas Sales Reports—Lifetime Totals—All Counties 7/31/2012," accessed November 1, 2012, aogc2.state.ar.us/FayettevilleShaleInfo/regularly%20 updated%20docs/B-43%20Field%20-%20Well%20MCF%20Totals.xls.

13. Southwestern Energy, "October 2012 Update," October 2012, swn.com /investors/LIP/latestinvestorpresentation.pdf, p. 23.

14. US Energy Information Administration, "Arkansas Natural Gas Marketed Production," released September 28, 2012, eia.gov/dnav/ng/hist/n9050ar2a .htm.

Chapter 17

1. Natural Resources Canada, "Energy Sector: Energy Sources: Canadian Natural Gas: Monthly Market Update: Historical Data," January 2001–April 2012, date modified 4/13/2011, nrcan.gc.ca/energy/sources/natural-gas /monthly-market-update/1173.

2. Baker Hughes, "North American Rotary Rig Count (January 2000–Current), Tab Canada Oil and Gas Split, October 26, 2012, investor.shareholder.com /common/download/download.cfm?companyid=BHI&fileid=608997& filekey=2349AA00-FC63-43FF-84F0-8E2432979DC3&filename=North _America_Rotary_Rig_Count_Jan_2000_-_Current_.xlsx.

3. Natural Resources Canada, "Energy Sector: Energy Sources: Canadian Natural Gas: Monthly Market Update: Historical Data," January 2001–April 2012, date modified 4/13/2011, nrcan.gc.ca/energy/sources/natural-gas /monthly-market-update/1173.

4. Christopher Adams, "The Status of Exploration and Development Activities in the Montney Play Region of Northeast BC," April 2, 2012, empr.gov .bc.ca/OG/oilandgas/petroleumgeology/UnconventionalGas/Documents /C%20Adams.pdf, slide 6.

5. BC Hydro, "Integrated Resource Plan Technical Advisory Committee Meeting #3," February 14, 2011, bchydro.com/etc/medialib/internet

/documents/planning_regulatory/iep_ltap/2011q1/irp_tac_mtg03_pre
sentation.Par.0001.File.IRP_TAC_Mtg03_Presentation_FNHRB_FINAL
_2011-02-11.pdf, slide 7.

6. National Energy Board, "Canadian Energy Overview 2011—Energy Briefing
Note," July 2012, neb-one.gc.ca/clf-nsi/rnrgynfmtn/nrgyrprt/nrgyvrvw
/cndnnrgyvrvw2011/cndnnrgyvrvw2011-eng.html, p. 10.

7. Ibid, p. 11.

8. Yara International, "Yara Expands Norway and Canada Fertilizer Capacity,"
news release, June 12, 2012, worldofchemicals.com/media/yara-expands
-norway-and-canada-fertilizer-capacity/3409.html.

9. National Energy Board, "Canadian Energy Overview 2011—Energy Briefing
Note," July 2012, neb-one.gc.ca/clf-nsi/rnrgynfmtn/nrgyrprt/nrgyvrvw
/cndnnrgyvrvw2011/cndnnrgyvrvw2011-eng.html, p. 11.

10. Ibid.

11. The Oil Sands Developers Group, "Oil Sands Project List," October 2012,
oilsandsdevelopers.ca/wp-content/uploads/2012/10/Oil-Sands-Project
-List-October-2012.pdf, p. 2.

Chapter 18

1. Federal Energy Regulation Commission, "North American LNG Import/
Export Terminals—Approved," July 17, 2012, ferc.gov/industries/gas/indus
-act/lng/LNG-approved.pdf.

2. Ibid.

3. Ibid.

4. bp.com, "BP Statistical Review of World Energy June 2012," accessed September 3, 2012, bp.com/sectionbodycopy.do?categoryId=7500&contentId
=7068481; bp.com/assets/bp_internet/globalbp/globalbp_uk_english
/reports_and_publications/statistical_energy_review_2011/STAGING
/local_assets/pdf/statistical_review_of_world_energy_full_report_2012
.pdf, p. 28.

5. Tamsin Carlisle, "Qatar Reaffirms North Field moratorium," *The National*,
March 9, 2009, thenational.ae/business/energy/qatar-reaffirms-north-field
-moratorium.

6. bp.com, "BP Statistical Review of World Energy June 2012," accessed September 3, 2012, bp.com/sectionbodycopy.do?categoryId=7500&contentId
=7068481; bp.com/assets/bp_internet/globalbp/globalbp_uk_english
/reports_and_publications/statistical_energy_review_2011/STAGING
/local_assets/pdf/statistical_review_of_world_energy_full_report_2012
.pdf, p. 28.

7. Ibid.

8. Reuters, "Royal Dutch Shell Plc (RDSa.L) Will Increase its Stake in
Australia's Browse LNG Project by Picking up Chevron's Equity in the
$30-billion Venture in an Asset-Swap Deal, Opening up the Possibility of

New Development Options Such as Floating LNG," August 21, 2012, reuters
.com/article/2012/08/21/us-shell-chevron-idUSBRE87K04E20120821.

9. Neil Hume, "Australia set to take its place on LNG stage," *Financial Times*,
April 24, 2012, ft.com/cms/s/0/3adce71e-8d9f-11e1-b8b2-00144feab49a
.html#axzz24aJCtmgF.

10. "EIU: Australia's coming LNG boom," *Oil & Gas Financial Journal*, August
8, 2012, ogfj.com/articles/2012/08/australias-coming-lng-boom.html.

11. bp.com, "BP Statistical Review of World Energy June 2012," accessed September 3, 2012, bp.com/sectionbodycopy.do?categoryId=7500&contentId
=7068481; bp.com/assets/bp_internet/globalbp/globalbp_uk_english
/reports_and_publications/statistical_energy_review_2011/STAGING
/local_assets/pdf/statistical_review_of_world_energy_full_report_2012
.pdf, p. 28.

12. bp.com, "BP Statistical Review of World Energy June 2012," accessed September 3, 2012, bp.com/sectionbodycopy.do?categoryId=7500&contentId
=7068481; bp.com/assets/bp_internet/globalbp/globalbp_uk_english
/reports_and_publications/statistical_energy_review_2011/STAGING
/local_assets/pdf/statistical_review_of_world_energy_full_report_2012
.pdf, p. 29 ; "Indonesia to Halt LNG Exports from Bontang, Arun Terminals by 2020," IHS.com, March 10, 2010, ihs.com/products/global-insight
/industry-economic-report.aspx?id=106594486.

13. bp.com, "BP Statistical Review of World Energy June 2012," accessed September 3, 2012, bp.com/sectionbodycopy.do?categoryId=7500&contentId
=7068481; bp.com/assets/bp_internet/globalbp/globalbp_uk_english
/reports_and_publications/statistical_energy_review_2011/STAGING
/local_assets/pdf/statistical_review_of_world_energy_full_report_2012
.pdf, p. 21.

14. bp.com, "BP Statistical Review of World Energy June 2012," accessed September 3, 2012, bp.com/sectionbodycopy.do?categoryId=7500&contentId
=7068481; bp.com/assets/bp_internet/globalbp/globalbp_uk_english
/reports_and_publications/statistical_energy_review_2011/STAGING
/local_assets/pdf/statistical_review_of_world_energy_full_report_2012
.pdf, p. 22.

15. US Energy Information Administration, "Carribean," May 1, 2012, eia.gov
/countries/regions-topics.cfm?fips=CR&trk=c.

16. US Energy Information Administration, "Trinidad and Tobago," May 1,
2012, eia.gov/countries/cab.cfm?fips=TD.

17. Ibid.

18. Government of the Republic of Trinidad and Tobago, "Review of the Economy 2011, From Steady Foundation to Economic Transformation," October
10, 2011, finance.gov.tt/content/Review-of-the-Economy-2011.pdf, page 15.

19. Ibid, p. 13.

20. Ibid.

21. Henning Gloystein and Jeff Coelho, "European slump leads utilities to burn more coal," news release, May 8, 2012, reuters.com/article/2012/05/08/us -energy-power-co-idUSBRE8470JZ20120508.

22. Ibid.

23. "Asia LNG Demand Robust Despite Soft Spot Market," 2012 South East Asia Australia Offshore Conference, August 10, 2012, seaaoc.com/news-old /asian-lng-demand-robust-despite-soft-spot-market, accessed October 24, 2012

24. Jasmine Wang and Kyunghee Park, "LNG Tankers Dodge Price Slump on China-Yard Shoutout: Freight," June 7, 2012, bloomberg.com/news/2012-06 -06/lng-tankers-dodge-price-slump-on-china-yard-shutout-freight.html; bp.com, "BP Statistical Review of World Energy June 2012," accessed September 3, 2012, bp.com/sectionbodycopy.do?categoryId=7500&contentId =7068481; bp.com/assets/bp_internet/globalbp/globalbp_uk_english /reports_and_publications/statistical_energy_review_2011/STAGING /local_assets/pdf/statistical_review_of_world_energy_full_report_2012 .pdf, p. 28.

25. bp.com, "BP Statistical Review of World Energy June 2012," accessed September 3, 2012, bp.com/sectionbodycopy.do?categoryId=7500&contentId =7068481; bp.com/assets/bp_internet/globalbp/globalbp_uk_english /reports_and_publications/statistical_energy_review_2011/STAGING /local_assets/pdf/statistical_review_of_world_energy_full_report_2012 .pdf, p. 29.

26. bp.com, "BP Statistical Review of World Energy June 2012," accessed September 3, 2012, bp.com/sectionbodycopy.do?categoryId=7500&contentId =7068481; bp.com/assets/bp_internet/globalbp/globalbp_uk_english /reports_and_publications/statistical_energy_review_2011/STAGING /local_assets/pdf/statistical_review_of_world_energy_full_report_2012 .pdf, p. 27.

27. ycharts.com, "Japan Liquefied Natural Gas Import Price Chart," accessed October 19, 2012, ycharts.com/indicators/japan_liquefied_natural_gas _import_price.

28. Yoree Koh, "Japan Restarts Reactor Amid Nuclear Protests," July 1, 2012, *Wall Street Journal*, online.wsj.com/article/SB10001424052702304299704577 500520506450482.html.

29. ycharts.com, "Japan Liquefied Natural Gas Import Price Chart," accessed October 19, 2012, ycharts.com/indicators/japan_liquefied_natural_gas _import_price.

30. Waterborne Energy, Inc. quoted in Federal Energy Regulatory Commission presentation, "World LNG Estimated August 2012 Landed Prices," July 11, 2012, google.com/url?sa=t&rct=j&q=&esrc=s&source=web&cd=1&ved=0

CCwQFjAA&url=http%3A%2F%2Fferc.gov%2Fmarket-oversight%2Fothr
-mkts%2Flng%2F2012%2F07-2012-othr-lng-archive.pdf&ei=j2hRUKPYB5
G89QSWoYHgAQ&usg=AFQjCNHmQTzP3Q3IUa9cuMEjHvmsHn8C
QA&cad=rja.

31. bp.com, "BP Statistical Review of World Energy June 2012," accessed September 3, 2012, bp.com/sectionbodycopy.do?categoryId=7500&contentId
=7068481; bp.com/assets/bp_internet/globalbp/globalbp_uk_english
/reports_and_publications/statistical_energy_review_2011/STAGING
/local_assets/pdf/statistical_review_of_world_energy_full_report_2012
.pdf, p. 29.

32. Platts, "Argentina Bets on Bolivian Gas as Cheaper Than LNG Imports,"
July 19, 2012, platts.com/RSSFeedDetailedNews/RSSFeed/NaturalGas
/8539396.

33. US Energy Information Administration, "U.S. Liquefied Natural Gas Imports," released September 28, 2012, eia.gov/dnav/ng/hist/n9103us2a.htm.

34. Ibid.

35. bp.com, "BP Statistical Review of World Energy June 2012," accessed September 3, 2012, bp.com/sectionbodycopy.do?categoryId=7500&contentId
=7068481; bp.com/assets/bp_internet/globalbp/globalbp_uk_english
/reports_and_publications/statistical_energy_review_2011/STAGING
/local_assets/pdf/statistical_review_of_world_energy_full_report_2012
.pdf, p. 28.

36. Federal Energy Regulatory Commission, "North American LNG Import/
Export Terminals—Existing," October 12, 2012, ferc.gov/industries/gas
/indus-act/lng/LNG-existing.pdf.

37. US Energy Information Administration, "U.S. Natural Gas Imports by
Country," release date September 28, 2012, eia.gov/dnav/ng/ng_move
_impc_s1_a.htm; US Energy Information Administration, "U.S. Natural
Gas Wellhead Price," released September 28, eia.gov/dnav/ng/hist/n9190
us3a.htm.

38. US Energy Information Administration, "U.S. Liquefied Natural Gas Imports," released September 28, 2012, eia.gov/dnav/ng/hist/n9103us2a.htm.

39. Robert Tuttle, "Mideast to Cut LNG Exports to Europe for First Time in
20 Years," June 1, 2012, dailystar.com.lb/Business/Middle-East/2012/Jun-01
/175325-mideast-to-cut-lng-exports-to-europe-for-first-time-in-20-years.ash
x#axzz28MEeq3mC.

40. Reuters, "UAE's Mubadala plans LNG terminal on Oman coast," news release, March 21, 2012, arabianbusiness.com/uae-s-mubadala-plans-lng-term
inal-on-oman-coast-450875.html.

41. Ibid.

42. Reuters, "Non-OPEC Producer Bahrain Plans to Award a Contract to
Build its Liquefied Natural Gas (LNG) Terminal by Year-End, Bahrain's

Energy Minister Said on Monday," news release, May 7, 2012, reuters.com /article/2012/05/07/bahrain-lng-idUSL5E8G737O20120507.

43. Robert Tuttle, "Mideast to Cut LNG Exports to Europe for First Time in 20 Years," June 1, 2012, dailystar.com.lb/Business/Middle-East/2012/Jun-01 /175325-mideast-to-cut-lng-exports-to-europe-for-first-time-in-20-years .ashx#axzz28MEeq3mC; Menafn.com, "Oman Will Probably Establish LNG Import Terminal," news release, December 10, 2011, menafn.com/men afn/1093449128/Oman-will-probably-establish-LNG-import-terminal.

44. 2b1stconsulting.com, "BP Scales Up to $24 Billion Tight Gas Oman Full Field Development," September 13, 2012, 2b1stconsulting.com/?s=bp+ scales+up+to+24+billion+Tight+Gas+Oman+Full+Field+Development/.

Chapter 19

1. US Energy Information Administration, "About EIA, Mission and Overview," accessed March 19, 2012, eia.gov/about/mission_overview.cfm.

2. Ibid.

3. Ibid.

4. Ibid.

5. Carolyn Cui, "Natural-Gas Data Overrated," *Wall Street Journal*, April 5, 2010, accessed March 19, 2012, online.wsj.com/article/SB1000142405270230 3912104575163891292354932.html.

6. Ibid.

7. US Energy Information Administration, "EIA-914 Monthly Gas Production Report Methodology Current as of April 2010," accessed March 19, 2012, eia .gov/oil_gas/natural_gas/data_publications/eia914/eia914meth.pdf, p. 5.

8. Ibid, p. 2.

9. Ibid, p. 4.

10. Texas Railroad Commission, "Monthly Summary of Texas Natural Gas— July 2012," updated October 17, 2012, rrc.state.tx.us/data/production/month lygas/2012/gasmonthlysummaryjuly12.pdf; US Energy Information Administration, "Monthly Natural Gas Gross Production Report," released September 2012, eia.gov/oil_gas/natural_gas/data_publications/eia914/eia 914.html.

11. US Energy Information Administration, "AEO 2011 Early Release Overview," 2011, eia/gov/forecasts/aeo/er/pdf/0383er(2011).pdf, p 1.

12. Ibid, p. 8.

13. Ibid, p. 1.

14. Intek, Inc., "Review of Emerging Resources: U.S. Shale Gas and Shale Oil Plays," December 2010, presented in US Energy Information Administration, "Review of Emerging Resources: US Shale Gas and Shale Oil Plays," July 2011, ftp.eia.doe.gov/natgas/usshaleplays.pdf.

15. Intek, Inc., "About Us," accessed November 9, 2012, inteki.com/about1.html.

16. Ian Urbina, "Behind Veneer, Doubt on Future of Natural Gas," June 6, 2011, nytimes.com/2011/06/27/us/27gas.html?pagewanted=all.
17. Ibid.
18. Intek, Inc., "Review of Emerging Resources: U.S. Shale Gas and Shale Oil Plays," December 2010, presented in US Energy Information Administration, "Review of Emerging Resources: U.S. Shale Gas and Shale Oil Plays," July 2011, ftp.eia.doe.gov/natgas/usshaleplays.pdf, p. 8.
19. Intek, Inc., "Review of Emerging Resources: U.S. Shale Gas and Shale Oil Plays," December 2010, presented in US Energy Information Administration, "Review of Emerging Resources: U.S. Shale Gas and Shale Oil Plays," July 2011, ftp.eia.doe.gov/natgas/usshaleplays.pdf, p. 5 of EIA portion of report.
20. Ibid.
21. Ibid.
22. Ibid.
23. Michigan Public Service Commission, "Michigan Natural Gas Production," accessed November 11, 2012, dleg.state.mi.us/mpsc/gas/pesec2.htm.
24. Ibid.
25. Intek, Inc., "Review of Emerging Resources: U.S. Shale Gas and Shale Oil Plays," December 2010, presented in US Energy Information Administration, "Review of Emerging Resources: U.S. Shale Gas and Shale Oil Plays," July 2011, ftp.eia.doe.gov/natgas/usshaleplays.pdf, p. viii of Intek portion of report.
26. Ibid.
27. Ibid.
28. Intek, Inc., "Review of Emerging Resources: U.S. Shale Gas and Shale Oil Plays," December 2010, presented in US Energy Information Administration, "Review of Emerging Resources: U.S. Shale Gas and Shale Oil Plays," July 2011, ftp.eia.doe.gov/natgas/usshaleplays.pdf, p. 6.
29. Ibid.
30. US Energy Information Administration, "AEO 2012 Early Release Overview," January 23, 2012, eia.gov/forecasts/aeo/er/pdf/0383er(2012).pdf, p 9.
31. Ibid.
32. Intek, Inc., "Review of Emerging Resources: U.S. Shale Gas and Shale Oil Plays," December 2010, presented in US Energy Information Administration, "Review of Emerging Resources: U.S. Shale Gas and Shale Oil Plays," July 2011, ftp.eia.doe.gov/natgas/usshaleplays.pdf, p. viii.
33. US Geological Survey, "USGS Releases New Assessment of Gas Resources in the Marcellus Shale," August 23, 2011, Appalachian Basin usgs.gov/newsroom/article.asp?ID=2893.
34. Jim Efstathiou, Jr. and Katarzyna Klimasinska, "U.S. to Slash Marcellus Shale Gas Estimate 80%," Bloomberg.com, August 23, 2011, bloomberg

.com/news/print/2011-08-23/u-s-to-slash-marcellus-shale-gas-estimate-80
-.html.

Chapter 20

1. Marc Roca and Ben Sills, "Solar Glut Worsens as Supply Surge Cuts Prices 93%: Commodities," Bloomberg.com, November 10, 2011, bloomberg.com /news/2011-11-10/solar-glut-to-worsen-after-prices-plunge-93-on-rising -supply-commodities.html.
2. Jeff Himmelman, "The Secret to Solar Powers," *New York Times*, August 9, 2012, nytimes.com/2012/08/12/magazine/the-secret-to-solar-power.html ?pagewanted=all.
3. First Solar, "First Solar Corporate Overview Q3 2011," accessed September 14, 2012, files.shareholder.com/downloads/FSLR/1395959378x0x477649/20 5c17cb-c816-4045-949f-700e7c1a109f/FSLR_CorpOverview.pdf, slide 14.
4. Ibid, slide 21.
5. Solar City, "SolarCity, Tesla, and UC Berkeley to Collaborate on Solar Storage Technologies," press release, September 20, 2010, solarcity.com/press releases/70/SolarCity-Tesla-and-UC-Berkeley-to-Collaborate-on-Solar -Storage-Technologies.aspx.
6. Liane Yvkoff, "BYD to provide energy storage solutions for LA," cnet.com, September 16, 2010, reviews.cnet.com/8301-13746_7-20016664-48.html.
7. Ibid.
8. Matthew Wald, "Batteries at Wind Farm Help Control Output," *New York Times*, October 28, 2011, nytimes.com/2011/10/29/science/earth/batteries -on-a-wind-farm-help-control-power-output.html?_r=2; Torresol Energy, "Central-Tower Technology," accessed September 14, 2012, torresolenergy .com/TORRESOL/central-tower-technology/en.
9. Business Wire, "AES Wind Generation and AES Energy Storage Announce Commercial Operation of Laurel Mountain Wind Facility Combining Energy Storage and Wind Generation," October 27, 2011, "businesswire.com /news/home/20111027006259/en/AES-Wind-Generation-AES-Energy -Storage-Announce.
10. Torresol Energy, "Central-Tower Technology," accessed September 14, 2012, torresolenergy.com/TORRESOL/central-tower-technology/en.
11. Torresol Energy, "Information about Gemasolar Plant," accessed November 6, 2011, torresolenergy.com/EPORTAL_DOCS/GENERAL/SENERV2 /DOC-cw4cb709fe34477/GEMASOLARPLANT.pdf.
12. Torresol Energy, "Gemasolar," accessed September 14, 2012, torresolenergy .com/TORRESOL/gemasolar-plant/en.
13. Ibid.
14. Torresol Energy, "Who We Are," accessed September 14, 2012, torresol energy.com/TORRESOL/who-we-are/en.

15. Torresol Energy, "Torresol Energy Launches Commercial Operations of Twin Solar Thermal Plants in Spain," news release, January 18, 2012, torre solenergy.com/TORRESOL/nota_prensa_detalle.html?id=cw4fi6b41 a2fb49.

16. US Department of Energy, "Secondary Energy Infobook (19 Activities)," 2009–2010, accessed November 7, 2012, www1.eere.energy.gov/education /pdfs/basics_secondaryenergyinfobook.pdf, pp. 59–60.

17. US Energy Information Administration, "Natural Gas Consumption by End Use," released November 2, 2012, eia.gov/dnav/ng/ng_cons_sum_dcu _nus_a.htm.

18. US Department of Energy, "Secondary Energy Infobook (19 Activities)," 2009–2010, accessed November 7, 2012, www1.eere.energy.gov/education /pdfs/basics_secondaryenergyinfobook.pdf, p. 58.

19. Ibid, p. 60.

20. Ministry of Economic Affairs, "Gas Production in the Netherlands, Importance and Policy," accessed November 7, 2012, sodm.nl/sites/default/files /redactie/gas_letter_eng.pdf, p. 5.

21. Ibid, p. 9.

22. Ibid, p. 5.

23. Ibid.

24. Ibid, p. 10.

25. Ibid, p. 10.

26. Laura De Angelo, "North Sea, Europe," eoearth.org, August 22, 2008, eoearth.org/article/North_Sea,_Europe.

27. East Texas Oil Museum.com, "A brief history of the East Texas Oil Field," accessed November 7, 2012, easttexasoilmuseum.com/Pages/history.html.

28. US Energy Information Administration, "Total Energy Annual Energy Review," release date September 27, 2012, eia.gov/totalenergy/data/annual /showtext.cfm?t=ptb0802a; United States Nuclear Regulatory Commission, "Power Reactors," March 29, 2012, nrc.gov/reactors/power.html.

29. Ibid.

30. US Energy Information Administration, "Table 7.2a Electricity Net Generation: Total (All Sectors)," release date September 26, 2012, eia.gov/total energy/data/monthly/pdf/sec7_5.pdf.

31. Tennesee Valley Authority, "TVA Releases Cost, Schedule Estimates for Watts Bar Nuclear Unit 2," news release, April 5, 2012, tva.com/news /releases/aprjun12/watts_bar.html.

32. World Nuclear Association, "Nuclear Power in the USA," October, 2012, accessed November 8, 2012, world-nuclear.org/info/inf41.html.

33. Katie Couric, "Al Gore: Energy Crisis Can Be Fixed," CBS Evening News, February 11, 2009, cbsnews.com/stories/2008/07/17/eveningnews/main 4270123.shtml; John Podesta and Timothy E. Wirth, "Natural Gas: A Bridge

Fuel for the 21stCentury," Center for American Progress, August 10, 2009, americanprogress.org/issues/green/report/2009/08/10/6513/natural-gas-a -bridge-fuel-for-the-21st-century/.

34. Bloom Energy, "ES-5700 Energy Server, Data Sheet," accessed November 7, 2012, bloomenergy.com/fuel-cell/es-5700-data-sheet/; BloomEnergy, "Solid Oxide Fuel Cells," accessed November 7, 2012, bloomenergy.com /fuel-cell/solid-oxide/.

35. Bloom Energy, "ES-5700 Energy Server, Data Sheet," accessed November 7, 2012, bloomenergy.com/fuel-cell/es-5700-data-sheet/.

36. Bloom Energy, "What is an Energy Server?," accessed November 7, 2012, bloomenergy.com/fuel-cell/energy-server/.

37. Bloom Energy, "Management Team, KR Sridhar, Ph.D," accessed November 7, 2012, bloomenergy.com/about/management-team/#sridhar.

38. Bloom Energy, "Reduce Energy Costs: Lower & Lock-In Energy Costs," accessed November 7, 2012, bloomenergy.com/reduce-energy-costs/.

39. Koch Financial Corp., "Koch Financial Corp. finances San Diego university's cogeneration facility," news release, October 30, 2001, power-eng.com /articles/2001/10/koch-financial-corp-finances-san-diego-universitys-co generation-facility.html.

40. Ucilia Wang, "Nuclear Startup NuScale finds a savior in Fluor," *Gigaom*, October 13, 2011, gigaom.com/cleantech/nuclear-startup-nuscale-finds-a -savior-in-fluor/.

41. TerraPower, "Home," accessed November 7, 2012, terrapower.com/home .aspx.

42. John Lippert and Jeremy van Loon, "Nuclear Scales Down," Bloomberg Markets, May 2001, p. 46.

43. World Nuclear News, "B&W, Bechtel Team Up on mPower," July 14, 2010, world-nuclear-news.org/NN-BandW_Bechtel_team_up_on_mPower-14 07106.html.

44. The Babcock & Wilcox Company, "B&M mPower Integrated System Test (IST) Facility," accessed November 8, 2012, babcock.com/products /modular_nuclear/ist.html.

45. Anthony York, "PG&E Spending on Proposition 16 Reaches $44 million," *Los Angeles Times*, May 24, 2010, latimesblogs.latimes.com/california -politics/2010/05/pge-spending-for-proposition-16-hits-44-million.html.

46. John Howard, "PG&E Price Tag for Prop. 16 Hits $44," *Capital Weekly*, May 24, 2010, capitolweekly.net/article.php?_c=109f9vsm908t1sg&1=1&xid =yv5lovwy2a1w6h&done=.yv5m7hsxrse7ig&_credir=1325788345&_c=109 f9vsm908t1sg.

47. Ibid.

48. California Public Utilities Commission, "Bundled Customer Rates by Class from 2000–2011," accessed November 8, 2012, ftp://ftp.cpuc.ca.gov/puc

/energy/electric/rates+and+tariffs/Average%20Rates%20by%20Customer
%20Class%20Years%202000-2011.ppt.

Chapter 21

1. Commodityonline.com, "Gas Producers Reduce Hedges in 2012 on Depressed Prices," accessed March 12, 2012, commodityonline.com/news/gas -producers-reduce-hedges-in-2012-on-depressed-prices-44885-3-1.html.
2. Ibid.
3. Encana Corporation, "Encana achieves 2011 operating targets 2012 capital investment focused on liquids exploration and development," news release, February 17, 2012, accessed March 12, 2012, encana.com/news-stories/news -releases/details.html?release=649543.
4. Ibid.
5. Chesapeake Energy Corporation, "Chesapeake Energy Corporation Reports Financial and Operational Results for the 2011 Fourth Quarter and Full Year," news release, February 12, 2012, accessed March 12, 2012, chk.com /news/articles/Pages/1663531.aspx.
6. Chesapeake Energy Corporation, "March 2012 Investor Presentation," accessed March 16, 2012, phx.corporate-ir.net/External.File?item=UGFyZW5 0SUQ9MTMwNDcofENoaWxkSUQ9LTF8VHlwZT0z&t=1, p. 39.
7. Baker Hughes, "North America Rotary Rig Counts—Tab U.S. Oil and Gas Split," accessed March 12, 2012, investor.shareholder.com/common/down load/download.cfm?companyid=BHI&fileid=552024&filekey=C9D38608 -D65C-40D2-9343-B4D99C8D4C80&filename=US_Rig_Report_030912 .xls.
8. AFP, "Troubled IKB Bank to Get 1.5-Billion-Euro Rescue Package: Minister," news release, February 13, 2008, afp.google.com/article/ALeqM5gfwBIUvl GjyqOSzRX3szCXv3V4Cg.
9. Paul J. Gough, "BG Group Cuts Rigs in Marcellus Shale," *Pittsburgh Business Times,* July 30, 2012, bizjournals.com/pittsburgh/blog/energy/2012/07/bg -group-cuts-rigs-in-marcellus-shale.html; Stanley Reed, "Series of Write-Downs Leads to a Loss at BP," *New York Times,* July 31, 2012, nytimes.com /2012/08/01/business/energy-environment/01iht-bp01.html; Neil Hume, "BHP Takes $2,84bn Writedown on Shale Gas," *Financial Times,* August 3, 2012, ft.com/cms/s/0/5df70b1c-ddoe-11e1-99f3-00144feab49a.html#axzz2 CPfok6hJ.
10. Aubrey McClendon, "Chesapeake Energy Corporation (CHK) Q3 2008 Business Update Call," October 15, 2008, 2:30 p.m. ET, transcript quoted in Seekingalpha.com, accessed July 26, 2011, seekingalpha.com/article /100644-chesapeake-energy-corporation-q3-2008-business-update-call -transcript; Ian Urbina, "Drilling Down: Documents: Leaked Industry

E-Mails and Reports," accessed September 7, 2011, nytimes.com/interactive/us/natural-gas-drilling-down-documents-4.html, p. 74.

11. Schlumberger Ltd., "Map of Austin Chalk Play," accessed March 15, 2012, glossary.oilfield.slb.com/en/Terms.aspx?LookIn=term%20name&filter=exploration.

12. Stephen Holditch and Husam AdDeen Madani, "Global Unconventional Gas — It Is There, But Is It Profitable?," JPT Online, accessed November 16, 2012, spe.org/jpt/print/archives/2010/12/11Management.pdf.

13. RedOrbit.com, "A Tale of Two Fields: Giddings Offers a Lesson in How Quickly Things Can Change," March 7, 2006, accessed March 15, 2012, redorbit.com/news/science/418617/a_tale_of_two_fields_giddings_offers_a_lesson_in/.

14. Ibid.

15. Texas Railroad Commission, "General Production Query Results, January 2011–December 2011, Field: Giddings (Austin Chalk-1), Well Type: Both, Monthly Totals," accessed November 4, 2012; Texas Railroad Commission, "General Production Query Results, January 2011–December 2011, Field: Giddings (Austin Chalk-3), Well Type: Both, Monthly Totals," accessed November 4, 2012.

List of Figures and Tables

271

Page   Tables

Bibliography

21b1stconsulting.com, "BP Scales Up to $24 billion Tight Gas Oman Full Field Development," September 13, 2012, 2b1stconsulting.com/bp-scales-up-to-24 -billion-tight-gas-oman-full-field-development/.

AAPG.org, "A 17-Year Overnight Sensation, Barnett Shale Play Going Strong," accessed August 29, 2011, aapg.org/explorer/2005/05may/barnett_shale.cfm.

Adams, Christopher, "The Status of Exploration and Development Activities in the Montney Play Region of Northeast BC," April 2, 2012, Montneyempr.gov .bc.ca/OG/oilandgas/petroleumgeology/UnconventionalGas/Documents /C%20Adams.pdf, slide 6.

AFP, "Troubled IKB Bank to Get 1.5-Billion-Euro Rescue Package: Minister," news release, February 13, 2008, afp.google.com/article/ALeqM5gfwBIUvlG jyqOSzRX3szCXv3V4Cg.

American Gas Association, "Our Country's Gas Supplies: What the Gas Industry is Doing to be Sure Your Home has Enough Gas," *Life* magazine, October 22, 1971, books.google.com/books?id=AEAEAAAAMBAJ&lpg=PA5&vq =American%20Gas%20association&pg=PA5#v=onepage&q&f=false.

American Gas Association, "Snapshot of U.S. Natural Gas Consumption (2008)," accessed May 17, 2011, aga.org/our-issues/issuesummaries/Pages /SnapshotUSNaturalGas.aspx, paragraphs 1 and 4.

Arkansas Oil and Gas Commission data, cited in Bill Powers, "A Brief History of a Shale Play," Powers Energy Investor, September 1, 2011.

Arkansas Oil and Gas Commission, "Fayetteville Shale Gas Sales Information, Lifetime Totals—All Counties 3/31/11," accessed June 28, 2011, aogc.state.ar.us /Fayprodinfo.htm.

Arkansas Oil and Gas Commission, "Fayetteville Shale Information—Well Gas Sales Reports—Lifetime Totals—All Counties 7/31/2012," accessed November 1, 2012, aogc2.state.ar.us/FayettevilleShaleInfo/regularly%20updated%20 docs/B-43%20Field%20-%20Well%20MCF%20Totals.xls.

"Asia LNG Demand Robust Despite Soft Spot Market," 2012 South East Asia Australia Offshore Conference, August 10, 2012, seaaoc.com/news-old/asian -lng-demand-robust-despite-soft-spot-market.

Association for the Study of Peak Oil & Gas—USA, "Wager Challenges CERA Oil Supply Prediction," February 6, 2008, accessed September 8, 2011, aspo -usa.com/archives/index.php?option=com_content&task=view&id=312& Itemid=91.

Baker Hughes Incorporated, "North America Rotary Rig Count Current Week Data/ US Oil/Gas Split," accessed May 19, 2011, investor.shareholder.com /bhi/rig_counts/rc_index.cfm.

Baker Hughes, "North America Rotary Rig Counts—Tab Canadian Oil and Gas Split," accessed June 7, 2011, investor.shareholder.com/bhi/rig_counts/rc_index.cfm.

Baker Hughes, "North America Rotary Rig Counts—Tab U.S. Oil and Gas Split," accessed March 12, 2012, investor.shareholder.com/common/download/download.cfm?companyid=BHI&fileid=552024&filekey=C9D38608-D65C-40D2-9343-B4D99C8D4C80&filename=US_Rig_Report_030912.xls.

Baker Hughes, "North America Rotary Rig Counts—Tab U.S. Oil and Gas Split," accessed June 7, 2011, investor.shareholder.com/bhi/rig_counts/rc_index.cfm.

Baker Hughes, "North America Rotary Rig Counts Current Week Data—Rigs by State—Current & Historical," September 7, 2012, investor.shareholder.com/common/download/download.cfm?companyid=BHI&fileid=598392&filekey=9CD5C2B7-07B1-451E-8385-B40D82FBD5BB&filename=Rigs_by_State_090712.xlsx.

Baker Hughes, "North America Rotary Rig Counts Through 2011," June 22, 2012, accessed September 29, 2012, investor.shareholder.com/common/download/download.cfm?companyid=BHI&fileid=579035&filekey=A3A9520E-DDE5-4682-ADBC-B384AC067849&filename=North_America_Rotary_Rig_Counts_through_2011.xls.

Baker Hughes, "North American Rotary Rig Count (January 2000–Current), Tab Canada Oil and Gas Split, October 26, 2012, investor.shareholder.com/common/download/download.cfm?companyid=BHI&fileid=608997&filekey=2349AA00-FC63-43FF-84F0-8E2432979DC3&filename=North_America_Rotary_Rig_Count_Jan_2000_-_Current_.xlsx.

Barnes, Tom, "2 drillers fined for Pennsylvania gas well blowout," *Pittsburgh Post-Gazette*, March 29, 2012, post-gazette.com/stories/local/state/2-drillers-fined-for-pennsylvania-gas-well-blowout-255250.

Barnett Shale Maps—Barnett Shale Maps, Specific Source unknown, accessed June 28, 2011, blumtexas.blogspot.com/2007/05/blog-post_190.html and 3.bp.blogspot.com/_tiLoSTplBIU/RstHX5S230I/AAAAAAAAAc4/ZE4J88LMdMg/s1600-h/bsmap51.jpg.

Basu, Kaustuv, "Fracking Open—Gas Drilling Research Stirs Controversies at Universities," *Inside Higher Ed* (July 6, 2012), accessed August 30, 2012, protecteaglesmere.org/2012/07/07/fracking-open/.

BBC News, "Fresh Alert over UK Gas Supplies—National Grid Has Issued its Latest 'Balancing Alert' on Gas Supplies," accessed June 27, 2011, news.bbc.co.uk/2/hi/business/8452805.stm.

BC Hydro, "Integrated Resource Plan Technical Advisory Committee Meeting #3," February 14, 2011, bchydro.com/etc/medialib/internet/documents/planning_regulatory/iep_ltap/2011q1/irp_tac_mtg03_presentation.Par.0001.File.IRP_TAC_Mtg03_Presentation_FNHRB_FINAL_2011-02-11.pdf, slide 7.

Ben P. Dell and Noam Lockshin, "Bernstein E&Ps: The Death Throes of the

Barnett Shale? Downgrading Devon to Market-Perform," Bernstein Research, May 13, 2010.

Berman, Arthur "ExxonMobil's Acquisition of XTO Energy: The Fallacy of the Manufacturing Model in Shale Plays" (*World Oil*, February 22, 2010), accessed May 20, 2011, theoildrum.com/node/6229.

Berman, Arthur E. and Lynn F. Pittinger, "U.S. Shale Gas: Less Abundance, Higher Cost," accessed, March 9, 2012, theoildrum.com/node/8212.

Berman, Arthur, Labyrinth Consulting Services, Inc., "Natural Gas Supply: Not as Great or as Inexpensive as Commonly Believed," American Association of Petroleum Geologists Conference, Houston Texas, April 12, 2011, accessed June 28, 2011, searchanddiscovery.com/.../2011/70097berman/ndx_berman .pdf.

Berman, Arthur, Labyrinth Consulting Services, Inc., "Shale Gas—Abundance or Mirage, Why the Marcellus Shale Will Disappoint Expectations," (ASPO USA 2010 World Oil Conference, Washington, D.C. October 8, 2010).

Berman, Arthur, Labyrinth Consulting Services, Inc., "Will The Plays Be Commercial? Impact on Natural Gas Price" (Middlefield Capital Presentation, July 2010), slide 8 and 10.

Bleizeffer, Dustin, "Gas Plant Still Idle After Fire," *Casper Star-Tribune*, June 9, 2010, trib.com/news/state-and-regional/article_ee198bda-27f7-54e4-a996-22 dfab1d54ea.html.

Bloom Energy, "ES-5700 Energy Server, Data Sheet," accessed November 7, 2012, bloomenergy.com/fuel-cell/es-5700-data-sheet/.

Bloom Energy, "Solid Oxide Fuel Cells," accessed November 7, 2012, bloom energy.com/fuel-cell/solid-oxide/.

Bogle, Jon to Dean (William) Easterling, attachment to Jon Bogle, "Penn State Admits Gas Study Flaws," July 1, 2010, accessed August 30, 2012, northcentral pa.com/article/penn-state-admits-gas-study-flaws.

Boonepickens.com, "T. Boone Pickens: His Life. His Legacy," accessed July 2, 2011, boonepickens.com/man_ahead/default.asp.

BP, "BP p.l.c Group Results, Second Quarter and Half Year 2012(a)," July 31, 2012, bp.com/liveassets/bp_internet/globalbp/STAGING/global_assets /downloads/B/bp_second_quarter_2012_results.pdf, p. 6.

bp.com, "BP Statistical Review of World Energy June 2011, Oil Section," accessed June 27, 2011, bp.com/sectiongenericarticle800.do?categoryId=903715 7&contentId=7068604, p. 6, 8.

bp.com, "BP Statistical Review of World Energy June 2011," bp.com/assets /bp_internet/globalbp/globalbp_uk_english/reports_and_publications /statistical_energy_review_2011/STAGING/local_assets/pdf/statistical _review_of_world_energy_full_report_2011.pdf, p. 22.

bp.com, "BP Statistical Review of World Energy June 2012," accessed August 23, 2012 and September 3, 2012, bp.com/assets/bp_internet/globalbp/globalbp _uk_english/reports_and_publications/statistical_energy_review_2011

/STAGING/local_assets/pdf/statistical_review_of_world_energy_full
_report_2012.pdf, p. 21, 22, 27, 28, 29.

Bryce, Robert, *Pipe Dreams* (New York, Public Affairs, 2002), 54.

Busby, Rebecca L., Ed, Institute of Gas Technology, *Natural Gas in Nontechnical Language*, (Tulsa, OK: PennWell, 1999), 95 and 98.

Business Wire, "AES Wind Generation and AES Energy Storage Announce Commercial Operation of Laurel Mountain Wind Facility Combining Energy Storage and Wind Generation," October 27, 2011, "businesswire.com /news/home/20111027006259/en/AES-Wind-Generation-AES-Energy -Storage-Announce

Business Wire, "Chesapeake Energy Corporation Announces Significant New Discovery in the Hogshooter Play of the Texas Panhandle and Western Oklahoma," news release, June 1, 2012, chk.com/News/Articles/Pages /1701619.aspx.

Business Wire, "Chesapeake Energy Corporation Comments on Inaccurate and Misleading *New York Times* Article," June 6/27/2011 5:52 AM,, accessed September 1, 2011, chk.com/news/articles/Pages/1579995.aspx.

Businessweek, "Plenty Of Oil—Just Drill Deeper," September 18, 2006, accessed July 11, 2011, businessweek.com/magazine/content/06_38/b4001055.htm.

Carlisle, Tamsin, "Qatar Reaffirms North Field moratorium," *The National*, March 9, 2009, thenational.ae/business/energy/qatar-reaffirms-north-field -moratorium.

Chakhmakhchev, Alex, Bob Fryklund, "Critical Success Factors of CBM Development—Implications of two strategies to global development," 19th World Petroleum Congress, Spain 2008, onepetro.org/mslib/servlet/one petropreview?id=WPC-19-3486.

Chesapeake Energy and BP, "Chesapeake and BP Announce Arkoma Basin Woodford Shale Transaction," news release, July 17, 2008, accessed July 26, 2011 chk.com/news/articles/pages/1176363.aspx.

Chesapeake Energy Corporation, "Chesapeake Energy Corporation Announces Sale of Fayetteville Shale Assets to BHP Billiton for $4.75 Billion in Cash," news release, February 21, 2011, chk.com/news/articles/pages/1530 960.aspx.

Chesapeake Energy Corporation, "Chesapeake Energy Corporation Provides Quarterly Operational Update," news release, August 2, 2010, chk.com/News /Articles/Pages/1455265.aspx.

Chesapeake Energy Corporation, "Chesapeake Energy Corporation Reports Financial and Operational Results for the 2012 Second Quarter," news release, August 6, 2012 11:24 am, chk.com/news/articles/Pages/1722883.aspx.

Chesapeake Energy Corporation, "Chesapeake Energy Corporation Reports Financial and Operational Results for the 2011 Fourth Quarter and Full Year," news release, February 12, 2012, accessed March 12, 2012, chk.com/news /articles/Pages/1663531.aspx.

Chesapeake Energy Corporation, "March 2012 Investor Presentation," accessed March 16, 2012, phx.corporate-ir.net/External.File?item=UGFyZW50SUQ9 MTMwNDcofENoaWxkSUQ9LTF8VHlwZToz&t=1, p. 39.

Chesapeake Energy, "2010 Institutional Investor and Analyst Meeting," Oklahoma City, OK, October 13, 2010, quoted in Ian Urbina, "Drilling Down— Insiders Sound an Alarm Amid a Natural Gas Rush," "Drilling Down, Documents: Leaked Industry E-mails and Reports," accessed November 13, 2011, nytimes.com/interactive/us/natural-gas-drilling-down-documents-4 .html p. 202.

Chesapeake Energy, "Chesapeake Energy Corporation Announces Sale of Fayetteville Shale Assets to BHP Billiton for $4.75 Billion in Cash," news release, February 21, 2011, accessed July 26, 2011, chk.com/news/articles/pages /1530960.aspx.

Chesapeake Energy, "Chesapeake Energy Corporation Provides Operational and Financial Update," news release, September 22, 2008, accessed July 27, 2011, chk.com/news/articles/Pages/1199524.aspx.

Chesapeake Energy, "Chesapeake Energy Corporation Reports Financial and Operational Results for the 2012 First Quarter," news release, May 1, 2012, accessed September 9, 2012, chk.com/news/articles/Pages/1689968.aspx.

Chesapeake Energy, 2011 Form 10-K, accessed August 22, 2012, phx.corporate-ir .net/External.File?item=UGFyZW50SUQ9MTMyNDcofENoaWxkSUQ9L TF8VHlwZToz&t=1, p. 16.

Cheseapeake Energy and BP, "Chesapeake and BP Announce Arkoma Basin Woodford Shale Transaction," news release, July 17, 2008, accessed July 26, 2011, chk.com/news/articles/pages/1176363.aspx.

"Chevron Acquires Marcellus Acreage from Chief Oil and Tug Hill. Cost May be Between $7,000-$11,000/acre!," May 5, 2011, accessed September 8, 2012, mergersandacquisitionreviewcom.blogspot.com/2011/05/chevron-acquires -marcellus-acreage-from.html.

"China's Labor Tests Its Muscle," *New York Times*, Updated August 16, 2010, accessed October 19, 2011, topics.nytimes.com/top/news/international/coun triesandterritories/china/labor-issues/index.html?scp=3&sq=minimum%20 wage%20hong%20kong&st=cse.

Clean Energy Fuels Corp., "Form 10-K" for period ended December 31, 2011, accessed August 30, 2012, investors.cleanenergyfuels.com/common/down load/sec.cfm?companyid=CLNE&fid=1047469-12-2470&cik=1368265, p. 35.

Cohen, David Michael, "Eagle Ford: Texas' Dark-Horse Resource Play Picks up Speed," *World Oil Online*, Vol. 232, No. 6, June 2011, accessed September 4, 2012, worldoil.com/June-2011-Eagle-Ford-Texas-dark-horse-resource-play -picks-up-speed.html.

Commodityonline.com, "Gas Producers Reduce Hedges in 2012 on Depressed Prices," accessed March 12, 2012, commodityonline.com/news/gasproducers -reduce-hedges-in-2012-ondepressed-prices-44885-3-1.html.

Congressional Board, Office of Technology Assessment, "An Analysis of the Impacts of the Projected Natural Gas Curtailments for the Winter 1975–76," NTIS order #PB2-50623 (November 1975), fas.org/ota/reports/7502.pdf, 12.

Congressional Budget Office, "Natural Gas Price Decontrol: A Comparison of Two Bills" (Congress of the United States, November, 1983), accessed, May 17, 2011, cbo.gov/doc.cfm?index=5076, pp. 3 and 4.

Considine, Timothy J., Robert Watson, Seth Blumsack, "The Economic Impacts of the Pennsylvania Marcellus Shale Natural Gas Play: An Update" (white paper, College of Earth & Mineral Sciences Department of Energy and Mineral Engineering, Penn State University), May 24, 2010, accessed August 30, 2012, marcelluscoalition.org/2010/05/the-economic-impacts-of -the-pennsylvania-marcellus-shale-natural-gas-play-an-update.

Considine, Timothy, Robert Watson, Rebecca Entler, Jeffrey Sparks, "An Emerging Giant: Prospects and Economic Impacts of Developing the Marcellus Shale Natural Gas Play" (white paper, College of Earth & Mineral Sciences Department of Energy and Mineral Engineering, Penn State University), July 24, 2009, accessed August 30, 2012 alleghenyconference.org /PDFs/PELMisc/PSUStudyMarcellusShale072409.pdf.

Cui, Carolyn, "Natural-Gas Data Overrated," Wall Street Journal, April 5, 2010, accessed March 19, 2012, online.wsj.com/article/SB10001424052702303912104575163891292354932.html.

Curtis, John, PhD, Chairman of the Potential Gas Committee, "Interview with Bill Powers," June 6, 2011, recorded with permission, duration 1:19:46. iTunes Voice Memo format.

De Angelo, Laura, "North Sea, Europe," eoearth.org, August 22, 2008, eoearth .org/article/North_Sea,_Europe.

De Bruin, Rod, email message to author, February 7, 2011.

Deffeyes, Kenneth S., Beyond Oil—The View from Hubbert's Peak (New York: Hill and Wang, a division of Farrar, Straus and Giroux, 2005), p. 79.

Delbridge, Rena, "No Guts, No Glory: John Martin's Hunches Have Paid Off in a Big Way for Wyoming," Made in Wyoming, accessed September 26, 2012, madeinwyoming.net/profiles/martin.php.

Dell, Ben, email message to author, June 20, 2011.

Department of Energy & Climate Change, "Fuel Used in Electricity Generation and Electricity Supplied (ET 5.1)," accessed June 27, 2011, decc.gov.uk/en /content/cms/statistics/energy_stats/source/electricity/electricity.aspx.

Department of Energy & Climate Change, "Natural Gas and Colliery Methane Production and Consumption, 1970 to 2009," accessed June 27, 2011, decc.gov .uk/en/content/cms/statistics/energy_stats/source/gas/gas.aspx.

Department of Energy & Climate Change, "Natural Gas Imports and Exports (Dukes 4.3)," accessed June 27, 2011, decc.gov.uk/en/content/cms/statistics /energy_stats/source/gas/gas.aspx.

Department of Energy & Climate Change, "Natural gas supply and consumption (ET 4.1), accessed June 27, 2011, decc.gov.uk/en/content/cms/statistics /energy_stats/source/gas/gas.aspx.

Department of Environmental Protection, "DEP Fines Chesapeake Appalachia $565,000 for Multiple Violations, news release, February 9, 2012, portal .state.pa.us/portal/server.pt/community/newsroom/14287?id=19258&type id=1.

Department of the Treasury, Bureau of the Mint, "Mint Facilities to Curtail Use of Natural Gas," press release, January 28, 1977, accessed May 17, 2011, usmint .gov/historianscorner/docs/pr514.pdf.

Devon Energy, Newsroom—News Release, "Devon Energy to Acquire Mitchell Energy for $3.5 Billion," accessed June 27, 2011, devonenergy.com/NEWS ROOM/Pages/NewsRelease.aspx?id=200032.

Doll, Thomas E., "Coalbed Methane 2009 Update Powder River Basin," Wyoming Oil and Gas Conservation Commission, July 6, 2009, deq.state.wy.us /out/downloads/Eggerman%20prbwater.pdf, p. 5.

Doveton, and Robert E. Barba, "Hugoton Asset Management Project (HAMP): Hugoton Geomodel Final Report" (Kansas Geological Survey) (2007).

Dube, H. G. et al. "SPE 63091: Lewis Shale, San Juan Basin: What We Know Now" (Society of Petroleum Engineers Inc., 2000).

East Texas Oil Museum.com, "A brief history of the East Texas Oil Field," accessed November 7, 2012, easttexasoilmuseum.com/Pages/history.html.

Econ.com, "Appendix C Natural Gas Regulation in the United States," (*Economic Insight, Inc. February 1996*), accessed July 12, 2011, econ.com/appc.pdf.

Efstathiou Jr., Jim, and Katarzyna Klimasinska, "U.S. to Slash Marcellus Shale Gas Estimate 80%," Bloomberg.com, August 23, 2011, bloomberg.com/news /print/2011-08-23/u-s-to-slash-marcellus-shale-gas-estimate-80-.html.

EH.net, "Manufactured and Natural Gas Industry," (Economic History Association, February 1, 2010), accessed May 17, 2011, eh.net/encyclopedia/article /castaneda.gas.industry.us, paragraph 18.

"EIU: Australia's coming LNG boom," *Oil & Gas Financial Journal*, August 8, 2012, ogfj.com/articles/2012/08/australias-coming-lng-boom.html.

Encana Corporation, "Encana achieves 2011 operating targets 2012 capital investment focused on liquids exploration and development," news release, February 17, 2012, accessed March 12, 2012, encana.com/news-stories/news -releases/details.html?release=649543.

"Encana Proposes Up to 3,500 New Wells Near Jonah Field," Wyoming Energy News, April 15, 2011, wyomingenergynews.com/2011/04/encana-proposes -up-to-3500-new-wells-near-jonah-field/.

Engelder, Terry, "Interview with Bill Powers," July 1, 2011, 1:02 pm, recorded with permission, duration 42:09, iTunes Voice Memo format; Terry

Engelder, "Interview with Bill Powers," July 6, 2011 1:09 pm, recorded with permission, duration 13.12, iTunes Voice Memo format.

Engelder, Terry, "Marcellus," *Fort Worth Oil and Gas Journal*, August, 2009, accessed September 8, 2011, www3.geosc.psu.edu/~jte2/references/references .html, link 155.

Englelder, Terry and Gary G. Lash, "Marcellus Shale Play's Vast Resource Potential Creating Stir In Appalachia," *American Oil & Gas Reporter*, May 2008, aogr.com/index.php/magazine/cover_story_archives/may_2008_cover _story/.

Environmental Defense Fund, "Clean Air Act Timeline—A Short History of Key Moments in One of the Most Effective Public Health Campaigns in U.S. History," accessed May 17, 2011 edf.org/documents/2695_cleanairact.htm.

Everett, Arthur, Associated Press, "Fierce Winter Cold Snap Causing Widespread Shortage of Oil, Gas," *The Lewiston (Maine) Daily Sun*, January 11, 1973, p. 6.

Exco Resources, Inc., "Exco Resources, Inc. Reports Second Quarter 2012 Results," news release, July 31, 2012, media.corporate-ir.net/media_files/irol/19 /195412/XCO_Q2_2012_PressRelease.pdf.

Exxon Mobil Corp, "10K Annual report pursuant to section 13 and 15(d)," Filed on 02/25/2011, Filed Period 12/31/2010, accessed July 25, 2011, ir.exxonmobil .com.

ExxonMobil, "ExxonMobil's Channel 100 year supply alt source," Format .flv, 5/19/2011 1:13 pm, duration: 0:31. accessed July 25, 2011, exxonmobil.com /Corporate/news.aspx.

Fassett, James E., "The San Juan Basin, a Complex Giant Gas Field, New Mexico and Colorado," *Search and Discovery*, #10254, August 31, 2010, searchand discovery.com/documents/2010/10254fassett/ndx_fassett.pdf, p. 33.

Federal Energy Regulation Commission, "North American LNG Import/Export Terminals—Approved," July 17, 2012, ferc.gov/industries/gas/indus-act /lng/LNG-approved.pdf.

Federal Energy Regulatory Commission, "North American LNG Import /Export Terminals—Existing," 10/12/12, ferc.gov/industries/gas/indus-act /lng/LNG-existing.pdf.

Federal Power Commissioner Rush Moody, Jr. to President Gerald Ford, 7 March 1975, The American Presidency Project, accessed May 17, 2011, presidency.ucsb.edu/ws/index.php?pid=4769&st=natural+gas&st1#axzz1 MjSbsAaV, paragraph 3.

FERC.gov, "North American LNG Import Terminals—Existing as of December 5, 2012," accessed December 8, 2012, ferc.gov/industries/gas/indus-act /lng/LNG-existing.pdf.

FERC.gov, "North American LNG Import/Export Terminals—Approved," July 17, 2012, ferc.gov/industries/gas/indus-act/lng/LNG-approved.pdf.

Fetete.com, F.A.S.T. CBM™ Coalbed Methane Reservoir Analysis, "Decline

Curve Analysis—Exponential, Hyperbolic, Harmonic Declines Illustration," accessed June 28, 2011, fekete.com/software/cbm/media/webhelp/c-te-tech niques.htm.

Fetkovich, M. J., E. J. Fetkovich and M. D. Fetkovich, Phillips Petroleum Co., "Useful Concepts for Decline-Curve Forecasting, Reserve Estimation, and Analysis" (Society of Petroleum Engineers Reservoir Engineering, February 1996), Number 00028628.

First Solar, "First Solar Corporate Overview Q3 2011," accessed September 14, 2012, files.shareholder.com/downloads/FSLR/.../FSLR_CorpOverview .pdf, slide 14 and 21.

Flores, R. M. and L. R. Bader, "A Summary of Tertiary Coal Resources of the Raton Basin, Colorado and New Mexico," U.S. Geological Survey Professional Paper 1625-A, pubs.usgs.gov/pp/p1625a/Chapters/SR.pdf.

FTC, Opinion 770-A, cited in Joseph P. Mulholland, Federal Trade Commission, "Economic Structure and Behavior in the Natural Gas Production Industry," February 1979.

Gargani, John C., "Dynamics of Starting a New Resource Play—Fayetteville Shale," November 18, 2008, plsx.com/finder/viewer.aspx?q=&f=src.100&v=0 &doc=6240&slide=127701#doc=6240&slide=127701&.

Geology.com, "Fayetteville Shale Orientation," accessed October 21, 2012, geology.com/articles/fayetteville-shale.shtml.

GlobalNewswire via Comtex News Network, "LINN Energy Announces 60.2 MMcfe Per Day Horizontal Granite Wash Well," news release, July 22, 2010, ir.linnenergy.com/releasedetail.cfm?ReleaseID=491038.

Gloystein, Henning and Jeff Coelho, "European slump leads utilities to burn more coal," news release May 8, 2012, reuters.com/article/2012/05/08/us -energy-power-co-idUSBRE8470JZ20120508.

Gough, Paul J., "BG Group Cuts Rigs in Marcellus Shale," *Pittsburgh Business Times*, July 30, 2012, bizjournals.com/pittsburgh/blog/energy/2012/07/bg -group-cuts-rigs-in-marcellus-shale.html.

Government of the Republic of Trinidad and Tobago, "Review of the Economy 2011, From Steady Foundation to Economic Transformation," October 10, 2011, finance.gov.tt/content/Review-of-the-Economy-2011.pdf, pp. 13, 15.

Govtrack.us, "H.R. 1835: New Alternative Transportation to Give Americans Solutions Act of 2009," accessed July 2, 2011, govtrack.us/congress/bill.xpd ?bill=h111-1835&tab=summary.

Granberg, Al, illustration in "What is Hydraulic Fracturing," ProPublica.org, accessed September 9, 2012, propublica.org/special/hydraulic-fracturing -national.

Harper, John and Jaime Kostelnik, "The Marcellus Shale Play in Pennsylvania," Pennsylvania Geological Survey, accessed September 8, 2012, marcellus.psu .edu/resources/PDFs/DCNR.pdf, p. 38.

Harper, John, "The Marcellus Shale—An Old 'New' Gas Reservoir in Pennsylvania," *Pennsylvania Geology*, Vol. 38, No. 1, Spring 2008, dcnr.state.pa.us /topogeo/pub/pageolmag/pdfs/v38n1.pdf, p. 3.

Himmelman, Jeff, "The Secret to Solar Powers," *New York Times*, August 9, 2012, nytimes.com/2012/08/12/magazine/the-secret-to-solar-power.html?page wanted=all.

History of Caddo Parish, "Oils and Natural Gas," accessed September 21, 2012, caddohistory.com/oil_gas.html.

Holditch, Stephen and Husam AdDeen Madani, "Global Unconventional Gas—It Is There, But Is It Profitable?," JPT Online, accessed November 16, 2012, jptonline.org/index.php?id=533.

Howard, John, "PG&E Price Tag for Prop. 16 Hits $44," *Capital Weekly*, May 24, 2010, millioncapitolweekly.net/article.php?_c=109f9vsm908t1sg&1=1&xid =yv5lovwy2a1w6h&done=.yv5m7hsxrse7ig&_credir=1325788345&_c=109f9 vsm908t1sg.

Huang, Wen-yuan, "Factors Contributing to the Recent Increase in U.S. Fertilizer Prices, 2002–08 (United States Department of Agriculture, February 2009), accessed October 17, 2011, ers.usda.gov/Publications/AR33/AR33.pdf, p. 4, 6 and 9.

Hume, Neil, "Australia set to take its place on LNG stage," *Financial Times*, April 24, 2012, ft.com/cms/s/0/3adce71e-8d9f-11e1-b8b2-00144feab49a.html#axzz 24aJCtmgF.

Hume, Neil, "BHP Takes $2,84bn Writedown on Shale Gas," *Financial Times*, August 3, 2012, ft.com/cms/s/0/5df70b1c-dd0e-11e1-99f3-00144feab49a.html #axzz2CPfok6hJ.

Hunter, J.K., "Abstract of: LaBarge Project: Availability of CO2 for Tertiary Projects," *Journal of Petroleum Technology*, Vol. 39, No. 11, (November 1987) onepetro.org/mslib/servlet/onepetropreview?id=00015160&soc=SPE.

"Indonesia to Halt LNG Exports from Bontang, Arun Terminals by 2020," IHS .com, March 10, 2010, ihs.com/products/global-insight/industry-economic -report.aspx?id=106594486.

Institute for 21st Century Energy, "1980s: Free Markets and the Decline of OPEC," accessed May 17, 2011, energyxxi.org/reports/1980s.pdf, p. 28.

Intek, Inc., "About Us," accessed November 9, 2012, inteki.com/about1.html.

Intek, Inc., "Review of Emerging Resources: U.S. Shale Gas and Shale Oil Plays," December 2010, presented in U.S. Energy Information Administration, "Review of Emerging Resources: U.S. Shale Gas and Shale Oil Plays," July 2011, ftp.eia.doe.gov/natgas/usshaleplays.pdf.

Intek, Inc., "Review of Emerging Resources: U.S. Shale Gas and Shale Oil Plays," December 2010, presented in U.S. Energy Information Administration, "Review of Emerging Resources: U.S. Shale Gas and Shale Oil Plays," July 2011, ftp.eia.doe.gov/natgas/usshaleplays.pdf, p. 5 of EIA portion of report, p. viii of Intek portion of report, 6, 8.

International Association for Natural Gas Vehicles, "Natural Gas Vehicle Statistics—NGV Count Ranked Numerically As at December 2010," accessed October 21, 2011, iangv.org/tools-resources/statistics.html.

Jackson, Robert and Avner Vengosh, "DEP: Protecting water or gas?" December 2, 2011, articles.philly.com/2011-12-02/news/30467569_1_drinking-water -water-resources-methane.

Jacobs, Andrew, "Honda Strikers in China Offered Less Than Demanded," *New York Times*, June 18, 2010, accessed October 19, 2011, nytimes.com/2010/06/19 /business/global/19strike.html?scp=10&sq=honda+foshan+china+strike& st=nyt.

Jeff Greene, Jeff, "The Greene Page, The Natural Gas Revival," April 5, 2011, accessed November 30, 2011, wisegasinc.com/wg-greene.htm.

Johnsson, Julie, "Nuclear Repairs No Easy Sale as Cheap Gas Hits Utilities," *Bloomberg.com*, September 11, 2012, bloomberg.com/news/2012-09-10/nuc lear-repairs-no-easy-sale-as-cheap-gas-hits-utilities.html

Kaiser, Mark J.and Yunke Yu, "Louisiana Haynesville Shale—1: Characteristics, Production Potential of Haynesville Shale Wells Described" (*Oil and Gas Journal*, December 5, 2011), accessed March 6, 2012, ogj.com/articles/print /volume-109/issue-49/exploration-development/louisiana-haynesville-shale -p1.html, p. 12.

Katie Couric, "Al Gore: Energy Crisis Can Be Fixed," CBS Evening News, February 11, 2009, cbsnews.com/stories/2008/07/17/eveningnews/main4270123 .shtml.

Koch Financial Corp., "Koch Financial Corp. finances San Diego university's cogeneration facility," news release, October 30, 2001, power-eng.com/artic les/2001/10/koch-financial-corp-finances-san-diego-universitys-cogenera tion-facility.html.

Koh, Yoree, "Japan Restarts Reactor Amid Nuclear Protests," July 1, 2012, *Wall Street Journal*, online.wsj.com/article/SB100014240527023042997045775005 20506450482.html.

Kostelnik, Jaime and Kristin M. Carter, "The Oriskany Sandstone Updip Permeability Pinchout: A Recipe for Gas Production in Northwestern Pennsylvania?," *Pennsylvania Geology*, Vol. 39, No. 4, Winter 2009, dcnr.state.pa.us /topogeo/pub/pageolmag/pdfs/v39n4.pdf, p. 20.

Lash, Eileen and Gary "Kicking Down the Well," (The SUNY Fredonial Shale Research Institute, The Early History of Natural Gas), accessed June 27, 2011, fredonia.edu/shaleinstitute/history.asp.

Lawrence, George H. "Bud", "Turnaround, Stories of: Behind the Scenes Political Washington, A Great American Industry and, Fun Along the Way," (Stillwater, OK: New Forums Press, Inc., 2007), p. 94.

Lazarus, David, "Don't Dismiss Pickens' Plan Yet," *Los Angeles Times*, July 9, 2008, accessed August 30, 2012, articles.latimes.com/print/2008/jul/09/busi ness/fi-lazarus9.

Lee, Don, "Battery Recharges Debate About U.S. Manufacturing," *Chicago Tribute*, May 16, 2010, accessed June 22, 2010, articles.chicagotribune.com/2010-05-16/business/ct-biz-0516-green-manufacture--20100516-5_1_battery-lithium-ion-cordless-power-tools.

Lereah, David, *Why the Real Estate Boom Will Not Bust and How You Can Profit From It* (New York: Random House, 2005).

Lippert, John and Jeremy van Loon, "Nuclear Scales Down," Bloomberg Markets, May 2001, p. 46.

Louisiana Bureau of Land Management, "Reasonably Foreseeable Develement Scenario for Fluid Minerals," March 2008, blm.gov/pgdata/etc/.../blm/...rf ds.../LA_RFDS_R2.pdf.

Louisiana Department of Natural Resources, "Haynesville Shale Wells Activity by Month," September 4, 2012, dnr.louisiana.gov/assets/OC/haynesville_shale/haynesville_monthly.pdf.

Louisiana Mid-Continent Oil and Gas Association, "History of the Industry—Oil and Gas 101," accessed September 21, 2012, lmoga.com/resources/oil-gas-101/history-of-the-industry.

Madhur, Smita, "Update 2-RLPC: Chesapeake increases bridge loan to $4B vs $3B, *Reuters*, May 15, 2012, accessed September 9, 2012, chk.com/news/articles/Pages/1689968.aspx.

Management and Effects of Coalbed Methane Produced Water in the United States, Committee on Management and Effects of Coalbed Methane Development and Produced Water in the Western United States, A Committee on Earth Resources, National Research Council (2010, Washington, DC: The National Academies Press), nap.edu/openbook.php?record_id= 12915, p. 93.

McClendon, Aubrey CEO, Chesapeake Energy Corporation, "Chesapeake Energy Corporation (CHK) *Chesapeake Energy* Corp Media *Conference Call*," July 11, 2011, recorded, duration 28:38, iTunes Voice Memo format.

McClendon, Aubrey K.', "Written Testimony of Aubrey K. McClendon, Chairman and CEO of Cheseapeake Energy Corporation and Chairman of the American Clean Skies Foundation, before The Select Committee on Energy Independence & Global Warming—July 30, 2008," accessed July 2, 2011, republicans.globalwarming.house.gov/.../2008/.../AubreyMcClendonTesti mony.pdf, p. 1.

McClendon, Aubrey, "Chesapeake Energy Corporation (CHK) Q3 2008 Business Update Call," October 15, 2008 2:30 PM ET, transcript, quoted in Seekingalpha.com, accessed July 26, 2011, seekingalpha.com/article/100644-chesapeake-energy-corporation-q3-2008-business-update-call-transcript; Ian Urbina, "Drilling Down: Documents: Leaked Industry E-Mails and Reports," accessed September 7, 2011, nytimes.com/interactive/us/natural-gas-drilling-down-documents-4.html, page 74.

McClendon, Aubrey, Bloomberg News, October 14, 2009, as quoted in Arthur

Berman, "Shale Gas—Abundance or Mirage? Why The Marcellus Shale Will Disappoint Expectations," theoildrum.com, October 28, 2010, 10:20 am, accessed June 28, 2011, theoildrum.com/node/7075.

McClendon, Aubrey, CEO, Chesapeake Energy Corporation, "Chesapeake Energy Corporation (CHK) Q2 2009 Earnings Call, Question-and-Answer -Session," August 4, 2009 9:00 AM ET seekingalpha.com/article/153691 -chesapeake-energy-corporation-q2-2009-earnings-call-transcript and seek ingalpha.com/article/153691-chesapeake-energy-corporation-q2-2009-earn ings-call-transcript?part=qanda.

McClendon, Aubrey, interview with Leslie Stahl. $halionaires (CBS "60 Minutes," November 14, 2010), FLV video, 13:25, youtube.com/watch?v=Vr6b -WzIcyo.

McRae, Gregroy S. and Carolyn Ruppel et al, "The Future of Natural Gas, an Interdisciplanary MIT Study" (MIT, June 2011), accessed July 25, 2011, web .mit.edu/mitei/research/studies/natural-gas-2011.shtml, page iii.

Menafn.com, "Oman Will Probably Establish LNG Import Terminal," news release, 12/10/2011, menafn.com/menafn/1093449128/Oman-will-probably -establish-LNG-import-terminal.

Methane Levels 17 Times Higher in Water Wells Near Hydrofracking," news release, Duke Nicholas School of the Environment, May 9, 2011, accessed September 9, 2012, nicholas.duke.edu/hydrofracking/methane-levels-17-times -higher-in-water-wells-near-hydrofracking-sites.

Michigan Public Service Commission, "Michigan Natural Gas Production," accessed November 11, 2012, dleg.state.mi.us/mpsc/gas/pesec2.htm.

Ministry of Economic Affairs, "Gas Production in the Netherlands, Importance and Policy," accessed November 7, 2012, sodm.nl/sites/default/files/redactie /gas_letter_eng.pdf, p. 5, 9, 10.

Moffett, James, McMoRan Exploration, The Oil & Gas Conference Presentation, August 15, 2011.

National Energy Board, "Canadian Energy Overview 2011—Energy Briefing Note," July 2012, neb-one.gc.ca/clf-nsi/rnrgynfmtn/nrgyrprt/nrgyvrvw/cnd nnrgyvrvw2011/cndnnrgyvrvw2011-eng.html, p. 10, 11.

Natural Gas Intelligence, "North American LNG Import Terminals," accessed June 8, 2011, intelligencepress.com/features/lng/.

Natural Gas Vehicle Institute, "NGV Connection Newsletter—August 2011," accessed November 30, 2011, ngvi.com/newsletter_august2011.html.

Natural Gas Vehicles for America, "Energy Policy Act of 1992—Fleet Program," accessed October 21, 2011, ngvc.org/gov_policy/fed_regs/fed_EPAFleetPrg .html.

Natural NGV Fleet Summit, "Nat Gas Act of 2009—New Alternative Transportation to Give Americans Solitions Act of 2009," accessed November 30, 2011, ngvsummit.com/natgasact.html.

Natural Resources Canada, "Canadian Natural Gas" Monthly Update October 2012, accessed December 11, 2012, nrcan.gc.ca/energy/sources/natural-gas /monthly-market-update/2258.

Natural Resources Canada, "Energy Sector: Energy Sources: Canadian Natural Gas: Monthly Market Update: Historical Data," January 2001–April 2012, date modified 4/13/2011, nrcan.gc.ca/energy/sources/natural-gas/monthly -market-update/1173.

Naturalgas.org, "History of Regulation—The Phillips Decision—Wellhead Price Regulation," accessed May 17, 2011, naturalgas.org/regulation/history .asp#phillips.

NaturalGas.org, "The History of Regulation—The Natural Gas Act of 1938," accessed May 17, 2011, naturalgas.org/regulation/history.asp#gasact1938.

Navigant Consulting Inc., "North American Natural Gas Supply Assessment, Prepared For American Clean Skies Foundation," July 4, 2008, accessed July 25, 2011, cleanskies.org/pdf/navigant-natural-gas-supply-0708.pdf.

New Mexico Bureau of Geology and Mineral Resources, "Coalbed Methane in New Mexico," *New Mexico Earth Matters*, Vol 4 No. 1, 2004, geoinfo.nmt.edu /publications/periodicals/earthmatters/4/EMV4N1.pdf.

New Mexico Oil Conservation Division, "Natural Gas and Oil Production [Friday, August 31, 2012]," accessed September 6, 2012, wwwapps.emnrd.state.nm .us/ocd/ocdpermitting/Reporting/Production/ProductionInjectionSumm aryReport.aspx.

Nordhaus, Robert R. "Producer Regulation and the Natural Policy Act of 1978," *Natural Resources Journal*, October 1979, Vol. 19, lawlibrary.unm.edu/nrj/19 /4/04_nordhaus_producer.pdf, p. 835.

North American Unconventional Gas Market Report 2007 (Warlick International, 2007), p. 163.

Oil & Gas Journal, "BHP Billiton Writes Down Fayetteville Shale Value," August 3, 2012, ogj.com/articles/2012/08/bhp-billiton-writes-down-us-shale-values .html.

Oil and Gas Journal, quoted in U.S. Energy Information Admministration, "Country Analysis Brief: Qatar," accessed June 22, 2011, eia.gov/countries/cab .cfm?fips=QA.

Olson, Laura, "Pa. fines driller $1.1 million over contamination, fire," *Pittsburgh Post-Gazette*, March 30, 2012, post-gazette.com/stories/local/breaking/pa -fines-driller-11-million-over-contamination-fire-298109/.

Orascom Cons. Inds., " Orascom Cons. Inds. (ORSD)—OCI Fertilizer Group Selects Iowa for New Plant," news release, September 5, 2012, bloomberg.com /article/2012-09-06/aK31H2deWv84.html.

Orascom Construction Industries, "OCI Beaumont Operation", accessed January 11, 2013, orascomci.com/index.php?id=pandoramethanolllc.

Osborna, Stephen G., Avner Vengoshb, Nathaniel R. Warnerb, and Robert B. Jackson, "Methane Contamination of Drinking Water Accompanying Gas-Well Drilling and Hydraulic Fracturing," Center on Global Change, Nicholas School of the Environment, Division of Earth and Ocean Sciences, Nicholas School of the Environment, and Biology Department, Duke University, Durham, NC 27708, January 13, 2011, nicholas.duke.edu/hydrofracking/resolve uid/eb4037ebd7c508e8a371f22bde2d4ac8, p. 2.

Pennsylvania Department of Environmental Protection, "Operator Active Wells," accessed July 5, 2011, dep.state.pa.us/dep/deputate/minres/oilgas /BOGM%20Website%20Pictures/2010/Operator%20Active%20Wells.jpg.

Pennsylvania Department of Environmental Protection, "Statewide Data Downloads by Reporting Period—Jul–Dec 2011 (Marcellus Only, 6 months)," accessed September 8, 2012, paoilandgasreporting.state.pa.us/publicreports /Modules/DataExports/ExportProductionData.aspx?PERIOD_ID=2011-2.

Pennsylvania Department of Environmental Protection, "Statewide Data Downloads by Reporting Period—Jan–Jun 2011 (Marcellus Only, 6 months)," accessed September 8, 2012, paoilandgasreporting.state.pa.us/publicreports /Modules/DataExports/ExportProductionData.aspx?PERIOD_ID=2011-1.

Pennsylvania Department of Environmental Protection, accessed May 12, 2011 and September 20, 2011, https://paoilandgasreporting.state.pa.us/public reports/Modules/Production/ProductionByCounty.aspx.

Pickens, T. Boone, "Interview with T. Boone Pickens on Squawkbox," April 28, 2011, accessed September 12, 2011, youtube.com/watch?v=vsC2SRs7cQE, duration 12:10, FLV480P format.

Pickens, T. Boone, "Lets Keep the Pressure On!," YouTube video 1:42 posted by "pickensplan," September 1, 2011, accessed November 30, 2011, youtube.com /user/pickensplan#p/u/8/7LscfDgArtE.

Pickens, T. Boone, "T. Boone Pickens Second TV Commercial," , August 1, 2008, accessed September 12, 2011, youtube.com/watch?v=X_3RV5SLS-I&feature =relmfu, duration 0.31, format FLV360P.

Pickensplan.com, email to author, "Please Sign My Petition Urging President Obama and Congress to Take Action on Energy Reform NOW!," January 8, 2010.

Pinedale Online, "Pinedale—Jonah Locator Map, Southwest, Wyoming," accessed September 26, 2012, pinedaleonline.com/news/2011/05/GasFields.htm.

Platts, "Argentina Bets on Bolivian Gas as Cheaper Than LNG Imports," July 19, 2012, platts.com/RSSFeedDetailedNews/RSSFeed/NaturalGas/8539396.

Podesta, John and Timothy E. Wirth, "Natural Gas: A Bridge Fuel for the 21st-Century," Center for American Progress, August 10, 2009, americanprogress .org/issues/green/report/2009/08/10/6513/natural-gas-a-bridge-fuel-for-the -21st-century/.

Potential Gas Committee, "Potential Supply of Natural Gas in the United States, Report of the Potential Gas Committee (December 31, 2010)," (Washington, DC, April 27, 2011), accessed July 2, 2011, potentialgas.org/ and PGC Press Conf 2011 slides.pdf linked from site, slide 8.

President Jimmy Carter remarks on the Emergency Natural Gas Act of 1977—remarks on signing S.474 and Related Documents, 2 February 1977, The American Presidency Project, accessed May 17, 2011, presidency.ucsb.edu/ws/index.php?pid=7422#axzz1SpT64e1, paragraph 4.

President Jimmy Carter to the Speaker of the House and the President of the Senate, 26 January 1977, The American Presidency Project, accessed May 17, 2011, presidency.ucsb.edu/ws/index.php?pid=7156#ixzz1MjZR2CKZ, paragraph 9.

President Richard Nixon to Congress, 18 April 1973. The American Presidency Project, accessed May 17, 2011, presidency.ucsb.edu/ws/index.php?pid=3817&st=&st1#axzz1MjSbsAaV, paragraph 35.

PRNewswire-USNewswire, "PA DEP Takes Aggressive Action Against Cabot Oil & Gas Corp. to Enforce Environmental Laws, Protect Public in Susquehanna County," news release, April 15, 2012, prnewswire.com/news-releases/pa-dep-takes-aggressive-action-against-cabot-oil--gas-corp-to-enforce-environmental-laws-protect-public-in-susquehanna-county-90951864.html.

PRNewswire, "Kinder Morgan-Copano Increase Presence in Eagle Ford Shale With New Long Term Contracts," June 30, 2011, accessed September 4, 2012, redorbit.com/news/business/2072981/kinder_morgancopano_increase_presence_in_eagle_ford_shale_with_new/.

"Producing Pinedale with a minimal footprint," Petroleum Economist, August 1, 2008, accessed on business.highbeam.com, accessed December 20, 2012, business.highbeam.com/435608/article-1G1-184291658/producing-pinedale-minimal-footprint.

Pyburn, Evelyn, "Denbury to Invest Billions in CO_2 Oil Recovery in Montana," Big Sky Business Journal, April 18, 2012, bigskybusiness.com/index.php/business/economy/2511-denbury-to-invest-billions-in-co2-oil-recovery-in-montana.

Range Resources Corporation,"Range Pioneered the Marcellus Shale Play in 2004 with the Successful Drilling of a Vertical Well, the Renz #1," accessed September 8, 2012, rangeresources.com/Operations/Marcellus-Division.aspx.

Range Resources, 2010 Annual Report, accessed September 4, 2011, phx.corporate-ir.net/phoenix.zhtml?c=101196&p=irol-reportsAnnual, page 43.

RedOrbit.com, "A Tale of Two Fields: Giddings Offers a Lesson in How Quickly Things Can Change," March 7, 2006, accessed March 15, 2012, redorbit.com/news/science/418617/a_tale_of_two_fields_giddings_offers_a_lesson_in/.

Reed, Stanley, "Series of Write-Downs Leads to a Loss at BP," New York Times,

July 31, 2012, nytimes.com/2012/08/01/business/energy-environment/01iht
-bp01.html.

Reginald Stuart, "Factories Widen Search For Winter's Fuel Supply," *The New York Times*, November 17, 1975.

Reginald Stuart, "Ohio, Starved for Natural Gas, Strives to Adjust Its Way of Life," *The New York Times*, February 3, 1977.

Report of the FTC to the U.S. Senate, S. Doc. No. 92, 70th Cong., 1st Sess. 588-91 (1936), 28 as reported in endnote n2, Pierce, Richard J. "Reconsidering the Roles of Regulation and Competition in the Natural Gas Industry" (*Harvard Law Review*, December 1983, 97 Harv. L. Rev. 345).

Reuters, "DAVOS-Shale gas is U.S. energy 'game changer' -BP CEO", January 28, 2010 7:43 am ET, accessed July 26, 2011, reuters.com/article/2010/01/28 /davos-energy-idUSLDE60R1MV20100128.

Reuters, "Non-OPEC Producer Bahrain Plans to Award a Contract to Build its Liquefied Natural Gas (LNG) Terminal by Year-End, Bahrain's Energy Minister Said on Monday," news release, May 7, 3012, reuters.com/article /2012/05/07/bahrain-lng-idUSL5E8G737O20120507.

Reuters, "Qatar to Hit Full LNG Export Capacity End 2011," June 5, 2011, accessed June 20, 2011, arabnews.com/economy/article448755.ece.

Reuters, "Royal Dutch Shell Plc (RDSa.L) Will Increase its Stake in Australia's Browse LNG Project by Picking up Chevron's Equity in the $30-billion Venture in an Asset-Swap Deal, Opening up the Possibility of New Development Options Such as Floating LNG," August 21, 2012, reuters.com/article /2012/08/21/us-shell-chevron-idUSBRE87K04E20120821.

Reuters, "UAE's Mubadala plans LNG terminal on Oman coast," news release, March 21, 2012, arabianbusiness.com/uae-s-mubadala-plans-lng-terminal-on -oman-coast-450875.html.

Rigby, Jon, CFA, "Global Nat Gas: When Will the Flood Ebb?" (UBS Investment Research Q-Seriesâ: Global Oil and Gas, February 26, 2010).

Roca, Marc and Ben Sills, "Solar Glut Worsens as Supply Surge Cuts Prices 93%: Commodities," Bloomberg.com, November 10, 2011, bloomberg.com /news/2011-11-10/solar-glut-to-worsen-after-prices-plunge-93-on-rising -supply-commodities.html.

Rubin, Jeff, *Why Your World Is About To Get a Whole Lot Smaller* (New York: Random House, 2009).

Schlumberger Ltd., "Map of Austin Chalk Play," accessed, March 15, 2012, glossary.oilfield.slb.com/DisplayImage.cfm?ID=156.

Sherrill, Robert, *The Oil Folies of 1970–1980* (New York: Anchor Press/Doubleday, 1983), 294.

Shirley, Kathy, "Country Has 'Abundant' Potential Report Tracks U.S. Gas Reserves," (Explorer), accessed June 8, 2011, aapg.org/explorer/2001/06jun /gas_update.cfm.

Shirley, Kathy, "Potential is Now a Reality—Coalbed Methane Comes of Age," *Explorer*, March 2000, accessed September 6, 2012, aapg.org/explorer/2000 /03mar/coalbedmeth.cfm.

"Shooters—a 'Fracking' History, American Oil & Gas Historical Society," accessed September 9, 2012, aoghs.org/technology/shooters-well-fracking -history/.

Shuster, Erik, "Tracking New Coal Fired Power Plants" (data update 1/13/2012), presentation U.S. Department of Energy, National Energy Technology Laboratory, January 13, 2012, 10.

Silver, Jonathan D., "The Marcellus Boom/Origins: the Story of a Professor, a Gas Driller and Wall Street," *Pittsburgh Post-Gazette*, March 20, 2011, accessed September 8, 2011, postgazette.com/pg/11079/1133325-503.stm.

Simons, Matt interviewed in Dave Cohen, "Questions About the World's Biggest Natural Gas Field," (*The Oil Drum*, June 9, 2006, 6:41 pm), accessed June 22, 2011, theoildrum.com/story/2006/6/8/155013/7696.

Sodersten, Sandra Sue and Debasish Sihi, Shell E&P Co.; and Marilyn Taggi Cisar, Shell EP-Americas, "HPHT, Low Permeability Fields in South Texas.... A Legacy of Development in a Tough Environment." Abstract. *International Petroleum Technology* Conference (Dubai, U.A.E., December 4–6, 2007), accessed March 15, 2012, onepetro.org/mslib/servlet/onepetro preview?id=IPTC-11581-ABSTRACT&soc=IPTC.

Solar City, "SolarCity, Tesla, and UC Berkeley to Collaborate on Solar Storage Technologies," press release, September 20, 2010, solarcity.com/pressreleases /70/SolarCity-Tesla-and-UC-Berkeley-to-Collaborate-on-Solar-Storage -Technologies.aspx.

Solorprices.org, "Benefits of Solar Carports: Green Living in the 21st Century," accessed September 3, 2012, solarprices.org/benefits-of-solar-carports-green -living-in-the-21st-century/#more-115.

Soraghan, Mike, "Natural Gas: U.S. Fracking Regulation Won't Halt 'Shale Gale'—report" (E&E Publishing LLC, March 10, 2010), accessed on June 22, 2011, eenews.net/public/eenewspm/2010/03/10/1.

Southwestern Energy, "October 2012 Update," October 2012, swn.com/investors /LIP/latestinvestorpresentation.pdf.

Southwestern Energy's CEO Discusses Q4 2010, Results—Earnings Call Transcript," posted on SeekingAlpha.com, February 25, 2011, seekingalpha.com/ article/255116-southwestern-energy-s-ceo-discusses-q4-2010-results-earnings -call-transcript?part=single.

Spencer, Starr, "Haynesville Shale Primed to Become World's Largest Gas Field by 2020," *Platts* republished on rigzone.com, February 11, 2009, rigzone.com /news/article.asp?a_id=72839.

Staaf, Erika, "Risky Business: An Analysis of Marcellus Shale Gas Drilling Violations in Pennsylvania 2008–2011," PennEnvironment Research & Policy

Center, February 2012, accessed September 9, 2012, pennenvironmentcenter
.org/sites/environment/files/reports/Risky%20Business%20Violations%20
Report_0.pdf

Standing, Tom, "Alaska's Key to Oil Production—It's a Gas…" (Association for
the Study of Peak Oil & Gas—USA, November 24, 2008), accessed May 17,
2011, aspousa.org/index.php/2008/11/alaskas-key-to-oil-production-its-a-gas,
paragraph 13.

State of Arkansas Oil and Gas Commission, "B-43 Field Well MCF (Totals
through 5/31/11 unless noted 8/25/11)," accessed August 29, 2011, aogc.state.ar
.us/Fayprodinfo.htm.

State of Louisiana, Department of Natural Resources, "Louisiana State Gas
Production, Wet After Lease Separation," revised June 4, 2012, accessed
September 17, 2012, dnr.louisiana.gov/assets/TAD/data/facts_and_figures
/table12.htm.

Statistics Canada. Table 131-0001—Supply and Disposition of Natural Gas,
Monthly (Cubic Metres), CANSIM (database), accessed May 6, 2011, 5.statcan
.gc.ca/cansim/a01?lang=eng.

Stewart, Robb M., "BHP Billiton's Petrohawk Swings to Lose," *The Wall Street
Journal—Market Watch*, August 9, 2012, marketwatch.com/story/bhp-billitons
-petrohawk-swings-to-loss-2012-08-09.

Stowers, Don, "Davy Jones discovery may herald new wave of drilling on GoM
shelf," *Oil & Gas Financial Journal*, March 1, 2010, ogfj.com/index/article-dis
play.articles.oil-gas-financial-journal.e-__p.Davy-Jones-discovery-may-her
ald-new-wave-of-drilling-on-GoM-shelf.QP129867.dcmp=rss.page=1.html.

Tennesee Valley Authority, "TVA Releases Cost, Schedule Estimates for Watts
Bar Nuclear Unit 2," news release, April 5, 2012, tva.com/news/releases/apr
jun12/watts_bar.html.

TerraPower, "Home," accessed November 7, 2012, terrapower.com/home.aspx.

Texas Railroad Commission, "Barnett Shale Information—Updated July 20,
2012" and "Newark, East (Barnett Shale) Statistics," accessed August 27, 2012,
rrc.state.tx.us/barnettshale/index.php.

Texas Railroad Commission, "Eagle Ford Fields and Counties 12-5-2012," accessed December 23, 2012, rrc.state.tx.us/eagleford/EagleFord_Fields_and
_Counties_201212.xls.

Texas Railroad Commission, "Eagle Ford Information," updated 08/30/12,
accessed August 31, 2012, rrc.state.tx.us/eagleford/index.php.

Texas Railroad Commission, "Eagle Ford Statistics Gas Production Statistics,"
Updated 12/18/2012, accessed December 22, 2012, rrc.state.tx.us/eagleford
/EagleFordGWGProduction.pdf.

Texas Railroad Commission, "General Production Query Results, January
2011–December 2011, Field: Giddings (Austin Chalk-1), Well Type: Both,
Monthly Totals," accessed November 4, 2012.

Texas Railroad Commission, "Monthly Oil and Gas Production by Year," 2006–2012, updated 7/25/2012, accessed September 12, 2012, rrc.state.tx.us/data/production/ogismcon.pdf.

Texas Railroad Commission, "Monthly Summary of Texas Natural Gas—July 2012," updated October 17, 2012, rrc.state.tx.us/data/production/monthlygas/2012/gasmonthlysummaryjuly12.pdf.

Texas Railroad Commission, "Natural Gas Production and Well Counts (1935–2011)," accessed September 12, 2012, rrc.state.tx.us/data/production/gaswell counts.php.

Texas Railroad Commission, "Oil and Gas Division District Boundaries," accessed October 21, 2011, rrc.state.tx.us/forms/maps/districts_colorsm.jpg.

Texas Railroad Commission, accessed September 20, 2011, rrc.state.tx.us/bossierplay/index.php.

The Babcock & Wilcox Company, "B&M mPower Integrated System Test (IST) Facility," accessed November 8, 2012, babcock.com/products/modular_nuclear/ist.html.

The Business Journal, "Pickens Backs Renewable Energy Plan," July 8, 2008, accessed July 2, 2011, bizjournals.com/triad/stories/2008/07/07/daily20.html.

The Economic Outlook: Before the Joint Economic Committee, U.S. Congress March 28, 2007 (Testimony of Chairman Ben S. Bernanke), accessed May 17, 2011, federalreserve.gov/newsevents/testimony/bernanke20070328a.htm, paragraph 6.

The Oil Drum blog, "Preditions for Canada's Natural Gas Production," blog entry by "benk," June 4, 2008 at 10:00 am, accessed November 30, 2011, theoildrum.com/node/4073.

The Oil Sands Developers Group, "Oil Sands Project List," October 2012, oilsandsdevelopers.ca/wp-content/uploads/2012/10/Oil-Sands-Project-List-October-2012.pdf, p. 2.

The Paleontological Research Institute, "Salt Dome Trap," accessed September 23, 2012, priweb.org/ed/pgws/systems/traps/structural/structural.html#salt.

Tippee, Bob, "Shale Gas supply expected to keep US prices low in 2011," Oil and Gas Journal, December 3, 2010, pennenergy.com/index/articles/pe-article-tools-template/_saveArticle/articles/oil-gas-journal/drilling-production-2/2010/12/shale-gas_supply_expected.html.

Torresol Energy, "Central-Tower Technology," accessed September 14, 2012, torresolenergy.com/TORRESOL/central-tower-technology/en.

Torresol Energy, "Gemasolar," accessed September 14, 2012, torresolenergy.com/TORRESOL/gemasolar-plant/en.

Torresol Energy, "Information about Gemasolar Plant," accessed November 6, 2011, torresolenergy.com/EPORTAL_DOCS/GENERAL/SENERV2/DOC-cw4cb709fe34477/GEMASOLARPLANT.pdf.

Torresol Energy, "Torresol Energy Launches Commercial Operations of Twin Solar Thermal Plants in Spain," news release, January 18, 2012, torresolenergy .com/TORRESOL/nota_prensa_detalle.html?id=cw4f16b41a2fb49.

Torresol Energy, "Who We Are," accessed September 14, 2012, torresolenergy .com/TORRESOL/who-we-are/en.

Tuttle, Robert, "Mideast to Cut LNG Exports to Europe for First Time in 20 Years," June 1, 2012, dailystar.com.lb/Business/Middle-East/2012/Jun-01/175 325-mideast-to-cut-lng-exports-to-europe-for-first-time-in-20-years.ashx#ax zz28MEeq3mC.

Two South Texas Pipelines to Handle Eagle Ford Shale Production," *Pipeline & Gas Journal*, February 2011, Vol. 238, No. 2, accessed September 4, 2012, pipelineandgasjournal.com/two-south-texas-pipelines-handle-eagle-ford -shale-production.

US Census, accessed May 12, 2011, quickfacts.census.gov/qfd/states/42/42015 .html.

US Department of Commerce, Bureau of Economic Analysis, "National Income and Product Accounts Table—Percent Change From Preceding Period in Real Gross Domestic Product," Accessed May 19, 2011, Years 1978–1987 selected, bea.gov/national/nipaweb/TableView.asp?SelectedTable=1&View Series=NO&Java=no&Request3Place=N&3Place=N&FromView=YES &Freq=Year&FirstYear=1978&LastYear=1987&3Place=N&Update=Update &JavaBox=no#Mid.

US Department of Energy, "Alternative Fuels Data center," accessed, January 11, 2013, afdc.energy.gov/fuels/electricity_locations.html.

US Department of Energy, "Energy Efficiency and Renewable Energy—Bringing You a Prosperous Future Where Energy is Clean, Abundant, Reliable and Affordable," October 2008, accessed October 21, 2011, apps1.eere.energy.gov /buildings/publications/.../bt_stateindustry.pdf, p. 5, 6 and 9.

US Department of Energy, "Natural Gas Wellhead Value and Marketed Production," 2010 Marketed Production, accessed December 15, 2012, eia.gov/dnav /ng/ng_prod_whv_dcu_nus_a.htm.

US Department of Energy, "Secondary Energy Infobook (19 Activities)," 2009–2010, accessed November 7, 2012, eere.energy.gov/.../pdfs/basics_secondary energyinfobook.pdf, pp. 59–60.

US Department of the Interior, Minerals Management Service, Gulf of Mexico OCS Region, "Outer Continental Shelf Estimated Oil and Gas Reserves, Gulf of Mexico," December 31, 2006, OCS Report MMS 2009-064, gomr.boe mre.gov/PDFs/2009/2009-064.pdf, p. v, 9.

US Department of Transportation Research and Innovative Technology Administration, "National Transportation Statistics," 2010, accessed September 3, 2012, bts.gov/publications/national_transportation_statistics/html/table _01_11.html.

US Energy Information Administation, "Qatar Accounts for a Growing Share of LNG Exports," accessed June 20, 2011, eia.gov/todayinenergy/detail.cfm ?id=50.

US Energy Information Administration, "AEO2011 Early Release Overview," 2011, 0383er(2011).pdf, p 1, 8.

US Energy Information Administration, "AEO2012 Early Release Overview," January 23, 2012, eia.gov/forecasts/aeo/er/pdf/0383er(2012).pdf, p. 9.

US Energy Information Administration, "Alaska South Field Production of Crude Oil," release date November 29, 2012, accessed December 10, 2012, eia.gov/dnav/pet/hist/LeafHandler.ashx?n=PET&s=MCRFPAKS1&f=M.

US Energy Information Administration, "Annual Energy Review 2009: Natural Gas Consumption by Sector, Selected Years, 1949-2009 (Billion Cubic Feet)," accessed May 19, 2011, eia.doe.gov/emeu/aer/pdf/pages/sec6_13.pdf.

US Energy Information Administration, "Annual U.S. Natural Gas Vehicle Fuel Consumption (Million Cubic Feet), release date 8/31/2012," accessed September 3, 2012, eia.gov/dnav/ng/hist/n3025us2A.htm.

US Energy Information Administration, "Arkansas Natural Gas Marketed Production," released September 28, 2012, eia.gov/dnav/ng/hist/n9050ar2a.htm.

US Energy Information Administration, "Carribean," May 1, 2012, eia.gov /countries/regions-topics.cfm?fips=CR&trk=c.

US Energy Information Administration, "Consumption for Electricity Generation by Energy Source: Total (All Sectors)," accessed May 19, 2011, eia.doe .gov/emeu/aer/txt/ptb0804a.html.

US Energy Information Administration, "Country Analysis Brief: Kuwait," accessed June 22, 2011, eia.gov/cabs/Kuwait/Full.html.

US Energy Information Administration, "Country Analysis Brief: Qatar," accessed June 22, 2011, eia.gov/countries/cab.cfm?fips=QA.

US Energy Information Administration, "Eagle Ford Shale Play, Western Gulf Basin, South Texas," map date: May 29, 2010, accessed October 24, 2011, eia.gov/oil_gas/rpd/shaleusa9.pdf.

US Energy Information Administration, "EIA-914 Monthly Gas Production Report Methodology Current as of April 2010," accessed March 19, 2012, eia.gov/oil_gas/natural_gas/data_publications/eia914/eia914meth.pdf, p. 2, 4, 5.

US Energy Information Administration, "Electricity Net Generation: Total (All Sectors), Selected Years, 1949–2009," accessed May 17, 2011, eia.doe.gov /emeu/aer/pdf/pages/sec8_8.pdf.

US Energy Information Administration, "Federal Offshore Gulf of Mexico, Natural Gas Gross Withdrawals and Production," released September 28, 2012, accessed September 28, 2012, eia.gov/dnav/ng/ng_prod_sum_dcu_r3 fm_m.htm.

US Energy Information Administration, "Figure 13.05 Wyoming Natural Gas

Marketed Production - Annual," release date 9/28/2012, accessed September 30, 2012, eia.gov/dnav/ng/hist/n9050wy2a.htm.

US Energy Information Administration, "Guidance1 for Federal Agencies: New Alternative Fuel Vehicle Definitions under Section 2862 of the National Defense Authorization Act of 2008," September 2008, accessed September 3, 2012, www1.eere.energy.gov/femp/pdfs/ndaa_guidance.pdf.

US Energy Information Administration, "Henry Hub Gulf Coast Natural Gas Spot Price (Dollars/Mil. BTUs)," accessed May 19, 2011, eia.gov/dnav/ng /hist/rngwhhdd.htm.

US Energy Information Administration, "Historical Natural Gas Annual 1930 through 2000," (1992–1996), December 2001, eia.gov/pub/oil_gas/natural _gas/data_publications/historical_natural_gas_annual/current/pdf/hn ga00.pdf.

US Energy Information Administration, "International Energy Statistics—Natural Gas Proved Reserves," accessed July 2, 2011, eia.gov/cfapps/ipdbproject /IEDIndex3.cfm?tid=3&pid=3&aid=6.

US Energy Information Administration, "Lower 48 States Shale Plays," (Energy Information Administration, May 9, 2011), accessed June 27, 2011, eia.gov/oil _gas/rpd/shale_gas.jpg.

US Energy Information Administration, "Monthly Natural Gas Gross Production Report," released September 2012, eia.gov/oil_gas/natural_gas/data _publications/eia914/eia914.html.

US Energy Information Administration, "Natural Gas 2006 Year-in-Review," accessed June 8, 2011, eia.gov/pub/oil_gas/natural_gas/feature_articles/2007/ ngyir2006/ngyir2006.pdf.

US Energy Information Administration, "Natural Gas Consumption by Sector, Selected Years 1949–2011 (Billion Cubic Feet)," Accessed December 10, 2012, eia.gov/totalenergy/data/annual/pdf/sec6_13.pdf.

US Energy Information Administration, "Natural Gas Consumption by End Use," released November 2, 2012, eia.gov/dnav/ng/ng_cons_sum_dcu_nus _a.htm.

US Energy Information Administration, "Natural Gas Deliveries to Commercial Consumers (Including Vehicle Fuel through 1996) in the US (Million Cubic Feet) release date 8/31/2012," accessed September 3, 2012, eia.gov/dnav /ng/hist/n3020us2A.htm.

US Energy Information Administration, "Natural Gas Gross Withdrawals and Production," released September 28, 2012, eia.gov/dnav/ng/ng_prod_sum _dcu_sar_a.htm.

US Energy Information Administration, "Natural Gas Wellhead Value and Marketed Production," released September 28, 2012, pulled for states: Texas: eia.gov/dnav/ng/ng_prod_whv_dcu_STX_a.htm; Louisiana: eia.gov/dnav /ng/ng_prod_whv_dcu_SLA_a.htm; Federal Offshore/Gulf of Mexico: eia

.gov/dnav/ng/hist/n9050fx2a.htm; Wyoming: eia.gov/dnav/ng/ng_prod
_whv_dcu_SWY_a.htm; New Mexico: eia.gov/dnav/ng/ng_prod_whv_dcu
_SNM_a.htm; Arkansas: eia.gov/dnav/ng/ng_prod_whv_dcu_SAR_a.htm.

US Energy Information Administration, "New Mexico Natural Gas Wellhead
Value and Marketed Production," release date 8/31/2012, accessed September
6, 2012, eia.gov/dnav/ng/xls/NG_PROD_WHV_DCU_SNM_A.xls.

US Energy Information Administration, "North Sea, Europe," republished by
eoearty.org, accessed August 29, 2011, eoearth.org/article/North_Sea,
_Europe, paragraph 3.1.2.

US Energy Information Administration, "Pennsylvania Natural Gas Marked
Production," released August 31, 2012, accessed September 8, 2012, eia.gov
/dnav/ng/hist/n9050pa2A.htm.

US Energy Information Administration, "Repeal of the Powerplant and Industrial Fuel Use Act (1987)," accessed May 19, 2011, eia.doe.gov/oil_gas/natural
_gas/analysis_publications/ngmajorleg/repeal.html.

US Energy Information Administration, "Shale Gas Production (Billion Cubic
Feet)," accessed June 27, 2011, eia.gov/dnav/ng/ng_prod_shalegas_s1_a.htm.

US Energy Information Administration, "Shale Gas Production (Billion Cubic
Feet)," accessed June 27, 2011, eia.gov/dnav/ng/ng_prod_shalegas_s1_a.htm;
Arkansas Oil and Gas Commission.

US Energy Information Administration, "Table 1.1: Net Generation by Energy
Source, Thousand Megawatthours: Total (All Sectors), Released Date, July
26, 2012, Data from Electric Power Monthly," accessed September 3, 2012,
eia.gov/electricity/monthly/xls/table_1_01.xlsx.

US Energy Information Administration, "Table 1.2: Existing Capacity by Energy Source, 2009 (Megawatts): Report," revised April 2011, accessed October 17, 2011, eia.gov/cneaf/electricity/epa/epat1p2.html.

US Energy Information Administration, "Table 7.2a Electricity Net Generation:
Total (All Sectors)," release date September 26, 2012, eia.gov/totalenergy
/data/monthly/pdf/sec7_5.pdf.

US Energy Information Administration, "Table 8.2a: Electricity Net Generation: Total (All Sectors), Selected Years, 1949–2009," accessed October 17,
2011, eia.gov/totalenergy/data/annual/txt/ptb0802a.html.

US Energy Information Administration, "Total Energy Annual Energy Review,"
release date September 27, 2012, eia.gov/totalenergy/data/annual/showtext
.cfm?t=ptb0802a.

US Energy Information Administration, "Trinidad and Tobago," May 1, 2012,
eia.gov/countries/cab.cfm?fips=TD.

US Energy Information Administration, "U. S. Natural Gas Industrial Consumption (Million Cubic Feet)," accessed October 17, 2011, eia.gov/dnav/ng
/hist/n3035us2a.htm.

US Energy Information Administration, "U.S. Liquefied Natural Gas Imports

(Million Cubic Feet)," accessed June 17, 2011, eia.gov/dnav/ng/hist/n9103 us2a.htm.

US Energy Information Administration, "U.S. Natural Gas Imports by Country," release date 9/28/2012, eia.gov/dnav/ng/ng_move_impc_s1_a.htm.

US Energy Information Administration, "U.S. Natural Gas Marketed Production (Million Cubic Feet), 1900–2011," accessed March 15, 2012, eia.gov/dnav /ng/hist/n9050us2a.htm.

US Energy Information Administration, "U.S. Natural Gas Marketed Production (Million Cubic Feet)," accessed May 17, 2011, May 19, 2012, eia.gov/dnav /ng/hist/n9050us2a.htm.

US Energy Information Administration, "U.S. Natural Gas Number of Gas and Gas Condensate Wells (Number of Elements)," accessed May 19, 2011, accessed June 7, 2011, eia.gov/dnav/ng/hist/na1170_nus_8a.htm.

US Energy Information Administration, "U.S. Natural Gas Pipeline Imports From Canada (Million Cubic Feet)," accessed May 19, 2011, June 7, 2011, eia.gov/dnav/ng/hist/n9102cn2a.htm.

US Energy Information Administration, "U.S. Natural Gas Residential Consumption (Million Cubic Feet), release date 8/31/12" accessed September 3, 2012, eia.gov/dnav/ng/hist/n3010us2A.htm.

US Energy Information Administration, "U.S. Natural Gas Total Consumption (Million Cubic Feet)" accessed May 17, 2011, eia.gov/dnav/ng/hist/n9140us 2A.htm.

US Energy Information Administration, "U.S. Natural Gas Wellhead Price (Dollars per Thousand Cubic Feet)—Annual Data," Release 11/30/2012, accessed December 11, 2012, eia.gov/dnav/ng/hist/n9190us3m.htm.

US Energy Information Administration, "U.S. Natural Gas Wellhead Price," released September 28, 2012, accessed September 29, 2012, eia.gov/dnav/ng /hist/n9190us3a.htm.

US Energy Information Administration, "U.S. Natural Gas Wellhead Price (Dollars per Thousand Cubic Feet)," accessed June 7, 2011, eia.gov/dnav/ng /hist/n9190us3a.html.

US Energy Information Administration, "US Coalbed Methane—Past, Present and Future," November 2007, accessed September 6, 2012, eia.gov/oil_gas /rpd/cbmusa2.pdf.

US Energy Information Administration, "What is Shale Gas and Why is it Important?" accessed June 27, 2011, eia.doe.gov/energy_in_brief/about_shale _gas.cfm.

US Energy Information Administration, "Wyoming Natural Gas Marketed Production—Annual data," released August 31, 2012, accessed September 26, 2012, eia.gov/dnav/ng/hist/n9050wy2a.htm.

US Energy Information Administration, Annual Energy Review "Table 8.4a: Consumption for Electricity Generation by Energy Source: Total (All

Sectors), Selected Years, 1949–2009," accessed October 17, 2011, eia.gov/total
energy/data/annual/txt/ptb0605.html.

US Energy Information Administration, Annual Energy Review, "Table 9.1:
Nuclear Generating Units, 1955–2009," accessed October 17, 2001, eia.gov
/totalenergy/data/annual/txt/ptb0901.html.

US Energy Information Administration, Federal Offshore—Gulf of Mexico
Natural Gas Marketed Production," (1997–2011), released September 28,
2012, accessed September 29, 2012, eia.gov/dnav/ng/hist/n9050fx2a.htm.

US Environmental Protection Agency, "Overview - The Clean Air Act Amendments of 1990," accessed May 19, 2011, epa.gov/oar/caa/caaa_overview.html.

US Environmental Protection Agency, "Reducing Acid Rain," accessed May 19,
2011, epa.gov/air/peg/acidrain.html.

US Geological Survey, "National Oil and Gas Assessment Project, 2010 Updated Assessment of Undiscovered Oil and Gas Resources of the National
Petroleum Reserve in Alaska (NPRA)," (Fact Sheet 2010–3102, October
2010), accessed July 2, 2011 pubs.usgs.gov/fs/2010/3102/.

US Geological Survey, "U.S. Geological Survey 2002 Petroleum Resource Assessment of the National Petroleum Reserve in Alaska (NPRA)," (Fact Sheet
045–02, 2002), accessed July 2, 2011, pubs.usgs.gov/fs/2002/fs045-02/.

US Geological Survey, "USGS Releases New Assessment of Gas Resources
in the Marcellus Shale," August 23, 2011, Appalachian Basin usgs.gov/news
room/article.asp?ID=2893.

US Geological Survey, map graphic used in Dustin Bleizeffer, "CBM: Bugs
vs. Bankruptcy," WyoFile.com, accessed September 28, 2012, wyofile.com
/2011/12/california-company-wyo-legislator-seek-to-delay-well-plugging
-with-hopes-that-microbes-will-rebuild-production/.

Ultra Petroleum, "About Us," accessed September 29, 2012, ultrapetroleum.
com/About-Us-4.html; Ultra Petroleum, "Geologic Setting," accessed September 29, 2012, ultrapetroleum.com/Our-Properties/Green-River-Basin
/Geologic-Setting-18.html.

Ultra Petroleum, "Location," accessed September 26, 2012, ultrapetroleum.com
/Our-Properties/Green-River-Basin/Location-17.html.

United States Nuclear Regulatory Commission, "Power Reactors," March 29,
2012, nrc.gov/reactors/power.html.

United States Nuclear Regulatory Commission, "Power Uprates for Nuclear
Plants," accessed October 17, 2011, nrc.gov/reading-rm/doc-collections/fact
-sheets/power-uprates.html.

Urbina, Ian, "Behind Veneer, Doubt on Future of Natural Gas," June 6, 2011,
nytimes.com/2011/06/27/us/27gas.html?pagewanted=all.

"US Shale and Australian Nickel Asset Review," BHP Billiton Limited, accessed
August 28, 2012, bhpbilliton.com/home/investors/news/Pages/Articles/US
-Shale-and-Australian-Nickel-Asset-Review.aspx.

"US Shale and Australian Nickel Asset Review," BHP Billiton Limited, accessed August 28, 2012, bhpbilliton.com/home/investors/news/Pages/Articles/US -Shale-and-Australian-Nickel-Asset-Review.aspx.

Wald, Matthew L., "Wisconsin Nuclear Reactor to Be Closed," *The New York Times*, October 22, 2012, nytimes.com/2012/10/23/business/energy-environ ment/dominion-to-close-wisconsin-nuclear-plant.html.

Wald, Matthew, "Batteries at Wind Farm Help Control Output," *New York Times*, October 28, 2011, nytimes.com/2011/10/29/science/earth/batteries -on-a-wind-farm-help-control-power-output.html?_r=2; Torresol Energy, "Central-Tower Technology," accessed September 14, 2012, torresolenergy .com/TORRESOL/central-tower-technology/en.

Wang, Jasmine and Kyunghee Park, "LNG Tankers Dodge Price Slump on China-Yard Shoutout: Freight," June 7, 2012, bloomberg.com/news/2012-06 -06/lng-tankers-dodge-price-slump-on-china-yard-shutout-freight.html.

Wang, Ucilia, "Nuclear Startup NuScale finds a savior in Fluor," *Gigaom*, Octo- ber 13, 2011, gigaom.com/cleantech/nuclear-startup-nuscale-finds-a-savior-in -fluor/.

Waterborne Energy, Inc. quoted in Federal Energy Regulatory Commission presentation, "World LNG Estimated August 2012 Landed Prices," July 11, 2012, google.com/url?sa=t&rct=j&q=&esrc=s&source=web&cd=1&ved=0C CwQFjAA&url=http%3A%2F%2Fferc.gov%2Fmarket-oversight%2Fothr-mk ts%2Flng%2F2012%2F07-2012-othr-lng-archive.pdf&ei=j2hRUKPYB5G89 QSWoYHgAQ&usg=AFQjCNHmQTzP3Q3IUa9cuMEjHvmsHn8CQA &cad=rja.

"What's Cooking with Gas: the Role of Natural Gas in Energy Independence and Global Warming Solutions," before the Select Committee on Energy Independence and Global Warming, 110th Cong. (July 30, 2006) (statement of Aubrey McClendon, CEO of Cheseapeake Energy), accessed July 2, 2011, globalwarming.house.gov/pubs/archives_110?id=0051#main_content and transcript: 110-46_2008-07-30.pdf, p. 13, 78 and 80.

White, Vince, Senior Vice President—Investor Relations, Devon Energy Cor- poration, "Interview with Bill Powers," May 24, 2011, 1:58 pm, Recorded with permission, duration 37:59. iTunes Voice Memo format.

Wicks, Frank, "The Oil Age," *Mechanical Engineering, the Magazine of ASME*, August 2009, memagazine.asme.org/Articles/2009/August/Oil_Age.cfm.

Williams, Peggy, "Prolific and Proud: A Century of E&P in Louisiana," *Oil and Gas Investor*A supplement, 2005, oilandgasinvestor.com/pdf/LASuplement .pdf, p. 11.

Williams, Peggy, "Wyoming—Operators in the Equality State are Driving Gas Production to New Highs, Thanks to Vigorous Drilling Programs in Both Old and New Areas," *Oil and Gas Investor*, March 2000, unitcorp.com/pdf /Wyoming.pdf, p. 32.

Wood, Robert E., "Earnings Skid for Top Three Steel Makers," *LA Times*, April 29, 1970.

World Nuclear Association, "Nuclear Power in the USA," October, 2012, accessed November 8, 2012, world-nuclear.org/info/inf41.html.

World Nuclear News, "B&W, Bechtel Team Pp on mPower," July 14, 2010, world-nuclear-news.org/NN-BandW_Bechtel_team_up_on_mPower-140 7106.html.

Wyoming Oil and Gas Conservation Commission, "Wyoming CBM Production MCF," accessed September 28, 2012, wogcc.state.wy.us/StateCbmGraph .cfm.

Wyoming Oil and Gas Conservation Commission, Production query system, "Report Date: 9/29/2012, Production for Year 2011 Based on Gas Production—Top 30 producing fields reviewed," wogcc.state.wy.us.

Wyoming Oil and Gas Conservation Commission, Production query system, "Report Date: 9/30/2012, Production for Year 1997 and 2011 Based on Gas Production—All fields reviewed," wogcc.state.wy.us.

Yara International, "Yara Expandes Norway and Canada Fertilizer Capacity," news release, June 12, 2012, worldofchemicals.com/media/yara-expands-nor way-and-canada-fertilizer-capacity/3409.html.

ycharts.com, "Japan Liquefied Natural Gas Import Price Chart," accessed October 19, 2012, ycharts.com/indicators/japan_liquefied_natural_gas_import _price.

Yergin, Daniel and Robert Ineson, "America's Natural Gas Revolution," (*Wall Street Journal*, November 2, 2009), accessed on 6/22/11, online.wsj.com /article/SB10001424052748703399204574507440795971268.html.

Yergin, Daniel, "It's Not the End Of the Oil Age," (*Washington Post*, July 31, 2005), accessed July 27, 2011, washingtonpost.com/wp-dyn/content/article /2005/07/29/AR2005072901672.html.

Yergin, Daniel, "Stepping on the Gas," (*Wall Street Journal*, April 2, 2011), accessed July 2, 2011, online.wsj.com/article/SB10001424052748703712504576 232582990089002.html.

York, Anthony, "PG&E Spending on Proposition 16 Reaches $44 million," *Los Angeles Times*, May 24, 2010, latimesblogs.latimes.com/california-politics /2010/05/pge-spending-for-proposition-16-hits-44-million.html.

Yvkoff, Liane, "BYD to provide energy storage solutions for LA," cnet.com, September 16, 2010, reviews.cnet.com/8301-13746_7-20016664-48.html.

Index

About the Author

BILL POWERS is an independent analyst
and private investor. Bill is the former editor of the Powers Energy Investor, the Canadian Energy Viewpoint and the US Energy
Investor. He has been publishing investment
research on the oil and gas industry since
2002 and sits on the Board of Directors of
Calgary-based Arsenal Energy. An active investor for over 25 years, Powers has devoted

Credit: Heidi Jo Brady

the last 15 years to studying and analyzing the energy sector, driven by
his desire to uncover superior investment opportunities.

If you have enjoyed *Cold, Hungry and in the Dark*
you might also enjoy other

BOOKS TO BUILD A NEW SOCIETY

Our books provide positive solutions for people who want to
make a difference. We specialize in:

Sustainable Living • Green Building • Peak Oil
Renewable Energy • Environment & Economy
Natural Building & Appropriate Technology
Progressive Leadership • Resistance and Community
Educational & Parenting Resources

New Society Publishers

ENVIRONMENTAL BENEFITS STATEMENT

New Society Publishers has chosen to produce this book on recycled paper made
with **100% post consumer waste,** processed chlorine free, and old growth free.
For every 5,000 books printed, New Society saves the following resources:[1]

33	Trees
2,958	Pounds of Solid Waste
3,255	Gallons of Water
4,246	Kilowatt Hours of Electricity
5,378	Pounds of Greenhouse Gases
23	Pounds of HAPs, VOCs, and AOX Combined
8	Cubic Yards of Landfill Space

[1]Environmental benefits are calculated based on research done by the Environmental Defense Fund
and other members of the Paper Task Force who study the environmental impacts of the paper
industry.

For a full list of NSP's titles, please call 1-800-567-6772 or check out our website at:

www.newsociety.com